# UFO

# FRIEND OR FOE?

Martin Thomas

10 9 8 7 6 5 4 3 2

This version has an index and a list of references, as well as some amendments.

## Dedication

This work is dedicated to the BLAZE Television Channel for making me think, to Google for helping me find information and to Wikipedia for denying everything.

## The Scientific Method

Develop an hypothesis which can be used to make testable predictions. Test these predictions and if they work, test again for repeatability.

## The Engineering Method

If it walks like a duck, and quacks like a duck, it probably is a duck.

**Table of Contents**

# PREFACE

Millions of people believe that there are mysterious objects flying in the sky. They were originally described as Unidentified Flying Objects – hence UFOs – but have more recently been termed UAPs – Unidentified Anomalous Phenomena. As most humans spend their time on the land, there are many more sightings of UAPs there, than have been associated with maritime UFOs, which have been termed Unidentified Submersible Objects (USOs). The abbreviation UAP is now used to cover both UFOs and USOs.

The term "Alien" has been used to describe any occupants or manufacturers of UAPs. However, this word has many pejorative associations – it means basically "not one of us" and is frequently used to describe immigrants, whose presence may be illegal, or who are disliked because of their race, colour or creed. I intend to use the word "Others", perhaps not an ideal term, but at least it avoids many of the pre-assumptions inherent in traditional terms.

Physicists, working in the field of String Theory, seem happy to contend with multiple universes, perhaps even an infinite number of them. It may be possible to move via "Portals" from one universe to another or, via "Gateways" into another universe and back into our universe again somewhere else. I am deliberately avoiding the term "Wormhole" which is another term for "Einstein-Rosen Bridge" which postulates links within our universe from one point to another, but which may be too dangerous to use anywhere near a planet.

Native American Shamans suggest that they can open Portals and communicate with the occupants of other universes, making use of vibrations – particularly drums and chanting. Shamans have been filmed inviting small UAPs to appear, and their consequent arrival, on the TV programme "Beyond Skinwalker Ranch". In the Indian sub-continent, the use of chanting appears to serve the same purpose, with mantras claimed to be very powerful.

For Gateways to work, the natures of Portals and Gateways may need to be fundamentally different. To transfer between universes may not require movement on the other side. To come out somewhere else in our universe may.

The obvious questions are: where have these UAPs come from, and what are they doing? It is not my intention to investigate the second of these questions, beyond where it may impact on the first one. An answer to this first question may assist in determining the answer to the second.

It must be remembered that we have no idea of the motivations of these "Others". They could be benign, helping our development as a species from pure altruism. They may want something from us or our planet, but are prepared to make deals with us. They may want our planet. Given the large number of "Other" species in the universe[i], all are possible at the same time. We have to hope some of these species are protecting us.

# *Chapter 1*

## IF THEY AREN'T "OTHER" IN ORIGIN, WHAT ARE UAPs ?

### a) GENERAL

UAPs seem to fall mainly into 2 distinct categories. They may be objects which are apparently metallic in construction, of many shapes and which can vary in size from larger than a car to being absolutely massive. Alternatively they can be balls, either of some form of plasma, or metallic, giving off or reflecting light, and varying from the size of a tennis ball upwards.

Let's look at the various explanations which have been put forward for where these UAPs come from, but which don't require physics which is currently beyond our comprehension:

- They don't exist
- They are hoaxes
- They are miss-interpreted natural phenomena
- Multiple reports are mass hysteria
- They are products of advanced earthly technology

Some governments make use of such explanations to refute possible UAP sightings, seeming to go out of their way to discredit witnesses, offering patently unscientific explanations, or ignoring inconvenient facts.

So, let's tackle these various proposals one by one.

## b) THEY DON'T EXIST

There are always people who chose to deny any evidence, even that of their own eyes. These Doubting Thomases don't feel the need to offer any explanation for what many people describe, and so don't deserve to have the benefit of being taken seriously themselves. Many governments appear to be determined to convince the general population that there is no such thing as a UAP, even to the extent of rubbishing sincere eye-witnesses.

## c) THEY ARE HOAXES

Undoubtedly there are many modern day hoaxes which do not survive proper examination. The TV series "The Proof is Out There" does a good job of testing these. Nevertheless, there are sightings which cannot be explained. Generally hoaxers relied on models and strings and, more recently, on doctored photographs and video recordings. However, multiple eye-witnessed events in the sky are sometimes difficult to explain.

There are also images from the past such as mediaeval paintings with apparent UAPs, such as "The Madonna with Saint Giovannino" attributed to Ghirlandaio, and "The Annunciation with Saint Emidius" by Carlo Crivelli". There is also a woodcut dated 1561 CE of a mass sighting of celestial objects over Nuremberg[2] which is difficult to ignore.

There are also anachronous artefacts found in genuine ancient archaeological strata. Examples might be The Antikythera[3] Mechanism, The Coso Artifact[4] and The Wedge of Alud[5].

Certainly, not all UAPs can be hoaxes.

## d) THEY ARE MISS-INTERPRETED NATURAL PHENOMENA

Many UAPs sightings are mistakes. Recording Venus as a UAP is reputedly a common error, and sightings of high-level American spy planes were apparently frequent (even if not a natural phenomenon). Lenticular[6] clouds, in particular, can be mistaken for massive "Mother Ships". It is theorised that balls of plasma can be generated by some earthquakes and ball lightening is an accepted short term phenomenon. These have all been government options for debunking reported sightings, with little respect for scientific credibility. The UK proposed the exact coincidence of a meteorite and an earthquake in a non-earthquake zone, to explain one sighting[7]. The USA once chose to explain a high-level ball of light as marsh gas[8].

These possible miss-interpretations do not mean that every UAP report is false. It has been estimated that, even with an honest debunking of many UAP reports, about 10% remain unexplained. By excluding sightings of solitary lights in the sky, I hope that this percentage is increased.

## e) MULTIPLE REPORTS ARE MASS HYSTERIA

This has been claimed for at least four multiple sightings by schoolchildren in Broadhaven[9] in Wales, Melbourne[10] in Australia, Miami[11] in the USA and Ruwa[12] in Zimbabwe. In all four cases, the witnesses, who are now adult, still stand by their descriptions of what they saw. It is also convenient for the nay-sayers to forget that their teachers were witnesses too. It has been reported that these teachers have been threatened with destruction of their careers if they don't keep silent.

The residents of Phoenix, Arizona would have been greatly offended if the sightings over their city in 1997 CE were described as mass hysteria, particularly as they were seen by an estimated 10,000 citizens. The event is described as the Phoenix Lights[13], and it was quickly explained away by the military as a flight of military aircraft and other aircraft dropping flares. The Arizona state governor, Fife Symington, responded with humour, introducing a "captured alien" who was actually his chief of staff, in an alien costume. However, once he had retired in 2007, he stated that he had acted in that way to prevent panic, and that he had actually seen the lights and, being an ex-pilot, knew they were neither conventional military aircraft nor flares.

## f) THEY ARE THE PRODUCTS OF ADVANCED EARTHLY TECHNOLOGY

Many of the witnesses of small aerial UAPs[14] have been highly qualified pilots, both commercial and military. Their testimonies have been uniformly the same – "We can't manoeuvre like that".

These UAPs are reported as showing instantaneous acceleration and deceleration to a far higher degree than human bodies could withstand, and sometimes jumping from one place to another. Also, if any nation had developed this technology, and it would have to be a major state to be able to afford it, its progressive development would have led to earlier applications which would probably have been used in a military context by now.

The larger UAPs seem to have been around for millennia: flying saucers of different sizes, big cylindrical vessels and massive "mother ships". They exceed the speed of sound without a sonic boom, and some have been seen to transition from air to water without apparently displacing water.

Circular UAPs have also been observed for millennia, ranging from 20 feet (7m) across, upwards to what have been called mother-ships. These could not be the product of current technology, as they easily pre-date it. However, in1967 CE, Jack Picket[15], a publisher of USA Air Force Publications, states that he was shown a range of circular aircraft at MacDill Air Force Base in Tampa which were dumped in their salvage yard. He was shown photographs of them in flight. These were up to 100 feet (30m) in diameter and were jet-powered. He was told that they had been very fast, and could fly very high, but had manoeuvrability problems.

There have been multiple reports of large triangular UAPs the size of football fields, in the Hudson Valley[16] in the USA in the 1980s, CE in the so-called Belgian Wave[17] in the early 1990s CE, in Bakewell[18] in the UK in 1993 CE, in the Phoenix Lights incident in 1997 CE, the Southern Illinois Incident[19] in 2000, the Tinley

Park Lights[20] in Illinois in 2004-6 CE, and the mass sighting at Stephensville Texas[21] in 2008 CE.

It is not clear whether these are truly triangular, or boomerang shaped, or perhaps both types exist, but they all exhibit light sources on the underside which appear to be the method of levitation and drive. Reports of these sightings suggest that they can hover and fly very rapidly, possibly turning by rotating on the spot, are effectively silent apart from a gentle hum, leave no discernible exhaust, and don't appear to work on current aerodynamic principles.

The location of these sightings points the finger of suspicion at the USA, but this doesn't explain how the technology was developed without some hint of what was going on. There certainly has been a determined attempt in the USA to discredit sightings and witnesses, with scary tales of thugs, sometimes called "Men in Black[22]", turning up to threaten some witnesses and even of sudden fatalities.

Many nations are developing drones technology, with perhaps the USA, Russia, Ukraine and China at the forefront. There are many reports of small bright lights and small solid objects. Many sightings appear to have been in East Asia and South America, with military videos of sightings on test ranges in the USA. There have been videos of small illuminated UAPs appearing and disappearing. It is not clear why Human military drones should even have the capability to be brightly illuminated, when invisibility would appear to the preferable.

Whilst modern drones can do amazing manoeuvres because they don't have pilots to limit the stresses they can take, it is doubtful whether even they can do everything reported, and there have been reports of highly manoeuvrable spherical UAPs from long before the electronic age when human drones first appeared.

In early South American cultures, there are detailed carvings of space beings[23] and spaceships. There are even early miniatures of aircraft[24] which have been shown experimentally to have credible aerodynamic capabilities. It is interesting that, in many places in South America, where ancient structures still stand, the local inhabitants claim that these buildings were already there when their ancestors first came to the region. Skulls[25] have been found in caves there, which are elongated at the back, similar to representations of Egyptian pharaohs such as Akhenaten. Mainstream archaeologists claim that these result from head-binding babies, but it has been shown that this does not enlarge the brain, whereas some of these ancient skulls have significantly larger brain volumes[26].

# *Chapter 2*

## WHAT ACTIVITIES ARE ATTRIBUTED TO THE "OTHERS"?

### a) PRE-HUMANITY

No-one seems interested in going back earlier than the age of the dinosaurs but, in some quarters, the "Others" get the blame for launching the asteroid which wiped them out. Unless they tell us that they did, we will never know.

Some geneticists believe that there is insufficient time between the extinction of the dinosaurs 66 million years ago, and the rise of humanity 20 thousand years ago, for natural selection to achieve what it did to humans. Other mammals in the fossil record appear to advance at a predictable rate, but humanity shows remarkable changes in the brain, which may have happened far too quickly. It is suggested there is a genetic discontinuity[27], which was engineered by the "Others".

## b) EARLY HUMANITY TO WORLD WAR II

Of all the achievements credited to the "Others", the most famous is the construction of the huge number of stone buildings from the Mesolithic period onwards, ranging from the stone circles in Orkney and Stonehenge in the UK, via Nan Madol[28] in the Pacific and the Floating Rock[29] in Japan, to Puma Puncu[xx] in Bolivia. In these places, and many more, there are carved stones so massive that it seems impossible that they were the handiwork of humans to carve them or move them. Many of these stones are granite, suggesting that the quartz crystals within could be used for some electrical or magnetic effect.

Early humans were hunter-gatherers until "Others" landed and became their leaders, teachers or kings. Egypt's pharaoh, Akhenaten[30], father of Tutankhamun, appears to have a misshapen head as did his wife and son, with it enlarged to the back and with other non-human differences. It is widely suggested that they were "Other" in origin. Similar misshapen skulls have been found in Malta and South America. Human parents, presumably wishing to make their children appear of royal background, have bound their babies' heads to achieve the same shape. However, it has been shown that they retain the same cranial volume as other children, whilst genuine misshaped skulls are much larger in volume.

There are records suggesting that the Sumarian civilisation was visited by winged "Others" which they called [31]Anunnaki. These beings were supposed to be humanoids, at least 4m tall, perhaps winged, and whose planet Nibiru is on a 3600 year orbit around our sun. They gave the early Sumerians agriculture, reading and writing, the wheel, and a code of law. Similar beings are represented on the walls of Egyptian buildings. The Anunnaki are sometimes shown accompanied by 2.5m tall humanoid beings with bird-like heads called Anzu in Sumerian legends[32], and Horus in Egypt[33]. These sound similar to bird-like "Others" claimed to live in the mountains of Japan today.

In the Book of Enoch, which is included in some versions of the Bible, he describes his being taken up into a spaceship where he meets many "Others". He is reputed to be the great grandfather of Noah.

China's first Emperor Qui Shi Huang[34] is thought to have either been "Other" or closely assisted by "Others". In Mexico, their Gods Quezalcoatl[35], Viracotcha[36] and many others are frequently described as the beings who came from the sky and taught the natives the basics of agriculture, medicine, mathematics and building.

In India, the "Others" appeared to have landed mob-handed, with spaceships called Vimanas[37], roughly conical UAPs, stylised representations of which are today a common feature in South East Asian architecture. In some early Indian drawings[38] they are shown as blue-skinned and humanoid in stature so they could be from Vega. They helped local humans but also squabbled amongst themselves. This squabbling culminated in the Kurushekta War[39] in about 1000 BCE, where Indian lore describes the gods flying round, and the use of devices to throw fire like lasers. Archaeologists have found the city Mohenjo

Daro[40] to have been completely flattened, and there are claims that it was covered in a layer of radioactive[41] ash, suggesting the use of a nuclear weapon.

Indian history suggests that they were in fact visited by two species of god – the Devas and the Asuras. These latter can be translated as Demons or Titans. Greek folklore describes a war between the Olympians and the Titans, won by the Olympians. The Olympians were human in appearance, whilst the Titans were giants. In India, the Titans are pictured as giants in their carvings, so they could be the same "Other" species. The winners write the history, so perhaps the Titans weren't quite as bad as they are pictured. Certainly, their presence would have assisted in many of the grand construction projects of those times.

Located in the in the fertile basin of the Upper Viliuy River, in Siberia, there is an area known to the nomad Siberian people as the Valley of Death. Here everything is so poisoned that nothing can live there. Across Siberia, giant metal structures may be found, hollow and half-buried in the soil of the forest. These have been nicknamed "Cauldrons[42]". Thanks to the remoteness of the region, their existence was thought to be a myth for many years. The entire area has always been shrouded with mystery, but a few facts emerge from history to tantalize modern scientists. The people of the area remember a legend that tells of a pillar of light. That light shone for many nights, and finally dimmed, only to flare again. It sent a fireball straight up into the air, which then sped across the land to strike down and obliterate the village's enemies. This occurred many times, sometimes in quick succession, and sometimes only after a wait of more than a hundred years.

Was this some form of world defence system? Are there similar "Cauldrons" elsewhere on earth, hidden in deserted areas such as the high Andes, Alaska and central Australia?

It is claimed that the "Others" either built or caused the building of pyramids, not just in Egypt, but everywhere in South and Central America, and in China and South East Asia. These are often associated with obelisks. It is suggested that these together once contributed to a world-wide power grid, with the Great Pyramid of Khufu being a prime source[43]. It is interesting to note that the obelisks and many menhirs are carved from granite with piezo-electric capabilities from the quartz in it.

The three pyramids at Giza are each aligned to due north, and it is claimed that they are aligned the same way as Orion's Belt[44]. The pyramids at Teotihuacan[45] in Mexico, specifically the Pyramid of the Moon, the Pyramid of the Sun, and the Pyramid of Quetzalcoatl[46], are often claimed to align with Orion's Belt.

The site of Göbekli Tepe[47] in Turkey has been proved to be over 11,000 years old, confounding archaeology's accepted time-line. The statues there show figures with six fingers.

There are now submerged coastal cities which show how the sea-level has risen since the end of the ice-age. As all the water held in the ice melted, it could have caused massive floods, and raised the sea level. Although the great flood, which is recorded in the Bible and other east Mediterranean texts, is generally assumed to be the Mediterranean Sea breaking into the Black Sea, there are records of a Great Flood in many cultures around the world. Some of these cultures are based in relatively high ground, and not all the melt-water in the world could produce floods at 3800m in the Andes or on a 1000m high plateau in Cappadocia[48]. It has been suggested that all the water came from a bombardment of ice-based planetessimals but, if there was sufficient water to cover all or most of earth, where did it all go?

It is suggested that there is evidence of flooding at Puma Puncu[49] in Bolivia in the Andes at a height of 12,000 feet, but this doesn't have to have come from a massive rise in sea level. It is very close to Lake Halnaymarca, and there may have been something to shake that lake at some time.

In 312 CE, the Emperor Constantine saw a cross in the sky before the battle of Milvian Bridge, and this convinced him to convert to Christianity[50]. It has been suggested that as this occurred in the vicinity of Monte Musine, it might have been an UAP.

The UK had its first recorded UAP sighting in 1113 CE[51], when pilgrims in the SW of England reported seeing a fire-belching dragon emerge from the sea, fly into the air and disappear into the sky.

In 1317 CE, a green circular UAP[52] hung over the Russian city of Tver, emitting a red glow and three rays. It was there for over a week before moving off.

At around dawn on 14 April 1561 CE, according to a broadsheet published that same year, "many men and women" of Nuremberg, Germany, saw what the broadsheet describes as an aerial battle with UAPs coming "out of the sun", followed by the appearance of a large black triangular object. Exhausted combatant spheres fell to earth in clouds of smoke. The broadsheet claims that witnesses observed hundreds of spheres, cylinders, and other odd-shaped objects that moved erratically overhead. The woodcut illustration depicts objects of various shapes, including crosses (with or without spheres on the arms), small spheres, two large crescents, a black spear, and cylindrical objects from which several small spheres emerged and darted around the sky.

In April 1665 CE, six fishermen claimed to witness an unexplained celestial phenomenon – an aerial battle in the skies above the Baltic Sea near Stralsund[53]. As evening broke, a dark-grey disk appeared high above the city centre. Great flocks of birds in the sky morphed into warships and engaged in a thunderous air battle. The decks teemed with ghostly figures. When, at dusk, "a flat, round shape like a plate" appeared above the local St. Nicholas Church, they fled. The following day, the fishermen found that they were trembling all over and complained of pains.

On June 30th 1908 CE, in Tunguska[54], Siberia, an oval-shaped mass swept across the sky and then there was a huge explosion. Locals reported that many trees were blown flat and more continued blazing for weeks. It was thought to have been a meteorite 200 feet in diameter, weighing some 100,000 tons. There were several expeditions to find it, between 1927 and 1939 CE, without finding the crash crater. Various theories are current: it was the crash, take-off or landing of a spaceship, it was a test of Tesla's secret energy weapon, or it was a signal from another species. No-one knows, but there are reports that some locals showed signs of genetic mutations similar to those caused by radiation. A further expedition in 1960 CE found quantities of nickel and iridium, suggesting there had been a mid-air explosion, and globules of melted dust suggested a nuclear event. An expedition in 2004 CE claimed to have found an "extraterrestrial device", but its nature has not been disclosed.

An UAP is reputed to have crash-landed in Italy in Magenta[55] in 1933 CE. It was supposedly examined by Marconi at the time, and taken away by the USA after the war.

In 1936 CE, an UAP allegedly crashed in the Black Forest near Frieberg[56], Germany and Hitler's SS quickly recovered it for research. It is suggested it was used as the basis of a Nazi flying saucer or Time Machine. However, there appears to be no evidence to support this until almost 30 years later.

## c) WORLD WAR II AND BEYOND

### i. General

During the Second World War, both Allied and Axis pilots reported what became known as Foo Fighters[57]. These were balls of light which were capable of out-manoeuvring their planes, and could interfere with their electronics if they got too close. There are even reports of their interfering with their engine electrical systems which were generally more low-tech than say radar systems. Allied pilots thought these were Axis secret weapons, and vice-versa.

There were several reports of larger flying saucers during the war, both in the USA and the UK. It is reported[58] that Winston Churchill and General Eisenhower discussed how to deal with the encounter, agreeing on a cover-up.

There are many reports of "Other" sightings during the large-scale NATO exercise Operation Mainbrace[59] in 1953. One of the participating warships was USS Eisenhower, which was the first USA ship to carry nuclear weapons, and was already no stranger to UAPs

Since World War II, the world has changed beyond measure, with human-produced rockets and the arrival of the Nuclear age, leading to a cold-war between USA and USSR. The resulting paranoia, particularly in the USA, must have led to fear of anything in the sky which couldn't be immediately explained. This would have worsened as mobile phone usage increased, and progressively more and more people became the owners of low-quality cameras. Also, the use of drones, both military and private, became more frequent. Nevertheless, not everything can be explained away as sightings of Venus, night-time drones, and experimental aircraft.

### ii. USA

There is a big difference between USA experiences and those of USSR/Russia as reported in the next section. Since the fall of Communism, official USSR government records have been made available to UAP researchers. Officially in the USA, UAPs do not exist and if anyone made a nuisance of themselves, they were ridiculed or threatened by government thugs – oops, sorry – Men in Black. Which is the totalitarian state here? It was only when some irrefutable USA navy video tapes were accidentally released that they had to admit that some UAPs were actually real.

During the war Germany, who had sent expeditions to Antarctica previously, sent a further expedition there in 1938-9 CE[60] despite all the other demands on its resources. It is suggested that this was connected to the crash-landed UAP in Germany in 1936 CE. It was thought that there was an "Other" base there. Straight after the war, when it is possible that the USA could have acquired both the German UAP and the Italian UAP, the USA launched Operation High Jump under the command of Admiral Byrd, to go to Antarctica. It is reported that Admiral Bird suddenly abandoned his expedition and hurried back the Washington after a confrontation with "Others".

When a massive NATO war-games exercise known as Operation Mainbrace convened in the North Sea in 1952 CE, it brought together 80,000 military personnel, 1,000 planes and 200 ships from nine countries. There were multiple sightings of UAPs which were documented by pilots and naval officers and appeared on radar.

From the initial detonation of a nuclear weapon on July 16 1945 CE onwards, there have been multiple reports of UAPs at test sites, nuclear facilities and ships carrying nuclear weapons. The most famous are possibly the Rendlesham Forest Incident at an USA air base in England on 25 December 1980 CE, and the USS

Nimitz film of 2004, where USA aircraft saw and videoed UAPs. When these videos came into public awareness, the USA government then acknowledged that they were genuine, but claimed that they didn't know what the objects were.

Back in the summer of 1952 CE, a fleet of UAPs is said to have buzzed Washington in the USA. The first incident occurred on 12 September when fireballs were sighted over a wide area of the eastern USA. It is claimed that an UAP impacted the earth at Flatwoods[61], West Virginia (300km from Washington), with reports of extensive air-to-air combat[62]. Sadly, an USA fighter pilot is recorded as dying that night.

Then, before midnight on July 20th, an air-traffic controller at what was then Washington National Airport reported seven unexplained radar blips[63], including over the White House and Capitol. Other controllers—at National and at Andrews Air Force Base—and even a pilot in the air, also reported strange objects. One controller at National, according to a 2002 Washington Post account, looked out of the tower and saw a "bright light hovering in the sky" that resembled "a saucer" and then took off at "incredible speed". Fighter jets were launched to intercept them. On the 22nd in Silver Spring, Maryland (10km from Washington), a piece of debris was discovered which it was claimed had been shot from an UAP. The next weekend, the 29th, the odd radar blips were back[64]—this time, air-traffic controllers counted a dozen. The government tried to claim they were all the result of a temperature inversion disrupting the radar. This might convince a layman but was an insult to the expertise of the air-traffic controllers themselves.

On Feb. 20, 1954 CE President Dwight Eisenhower is claimed[65] to have interrupted his vacation in Palm Springs, Calif., to make a secret nocturnal trip to Edwards Air Force base, where he met with two ETs with white hair, pale blue eyes and colourless lips. These ETs are nicknamed "Nordics" in UAP circles because they resemble Scandinavian humans. They are said to come from Rigel and Procyon, and are slightly taller than humans. The "Nordics" offered to share their superior technology and their spiritual wisdom with Ike if he would agree to eliminate America's nuclear weapons. Ike declined the ETs' offer, because he did not want to give up the nukes unilaterally.

Sometime later in 1954 CE, Ike reached a deal with another race of ETs, known as the "Greys" [66], allowing a permanent base in the USA, and permitting them to capture earthling cattle and humans for medical experiments, provided that they returned the humans safely home. "The Greys" is a convenient way to describe two grey-skinned species who often work together. These are imaginatively called the Tall Greys and Small Greys. They both supposedly come from the southern constellation Reticulum, and the Small Greys generally take the lead. The Tall Greys are thought to be a less-evolved version of the species. The Small Greys are the most observed "Others" on our planet.

The Small Greys are short, at about 1.0m - 1.3m tall, with a spindly body and legs and a large head with very big eyes. The Tall Greys are typically human height but with a larger head and eyes.

There appear to have been a number of UAP crashes in the USA since 1945[67]. Some of the early ones were The Trinity Incident[68]

1945 CE, Roswell[69] 1947 CE, and Kingman[70] 1953 CE. In one of these, it is reported that an "Other" called J Rod[71], was found alive, and it contributed significantly to the USA's attempts at reverse-engineering flying saucers before it died some years later. Bob Lazar[72], a whistle blower in 1989 CE, said that the USA held nine flying saucers which they had salvaged from crashes, and which they were reverse engineering.

What is difficult to understand is how a civilisation with technical ability to fly across the galaxy, can produce such terrible flying saucers that they keep falling out of the sky. This makes one wonder who was building these later versions – were some of these USA prototypes similar to those that had been seen by Jack Pickett in a salvage yard at MacDill Air Force Base in 1967 CE?

In December 1965 CE, thousands saw an UAP leave a streak of fire across the sky, change direction and crash in woods near Kecksburg[73], Pennsylvania. Witnesses there described the machine as roughly acorn-shaped. They described how armed military turned up and took it away. Much later, it was claimed that the UAP looked like Die Glocke (The Bell), an UAP which was developed in Nazi Germany in 1945 CE. The question is, of course, which was the chicken, and which was the egg?

Many people have claimed to have been abducted by "Others". They describe being subjected to medical tests, sometimes interfered with sexually, and sometimes finding some form of implant in their body afterwards. It is not known whether all abductees are returned.

There are many reports of animal mutilations, with horses and cattle being killed, drained of their blood and reproductive and facial organs removed. These mutilations are often correlated with the appearance of UAPs and, sometimes, the dead animal appears to have been dropped from a height afterwards.

There have been multiple reports of large triangular UAPs the size of football fields, in the Hudson Valley in the USA in the 1980s CE, in the so-called Belgian Wave in the early 1990s, in Bakewell in the UK in 1993 CE, Russia in 1989-1990 CE, in the Phoenix Lights incident in 1997 CE, the Southern Illinois Incident in 2000 CE, the Tinley Park Lights in Illinois in 2004-6 CE, and the mass sighting at Stephensville, Texas in 2008 CE.

There is believed to be a sprawling underground complex located beneath Archuleta Mesa on the Jicarilla Apache Reservation near Dulce[74], New Mexico. Dulce Base is said to contain at least eight levels that extend up to two miles below the surface. According to reports, the bottom three are staffed by as many as 18,000 Greys, as well as members of other species such as Reptoids – the "Others" base as agreed by Eisenhower? There are stories of a gunfight there in 2021 CE, although no-one seems to know why, nor the outcome. It is also claimed there are tunnels for high-speed shuttle trains to connect Dulce to other similar sites across the western USA and that there are landing facilities for aircraft using UAP technology.

In 1964, a UAP was filmed during a US missile test[75] firing, attacking and destroying it over the Pacific Ocean. On March 27 1967 CE, a UAP flew over the Malmstrom Air Force Base in Montana, which housed ICBM silos, and this was followed by ten nuclear weapons going off-line for a short period. This may have been the Greys acting on their hidden agenda to stop all military nuclear activities.

On 17th November 1986 CE, Japanese Airlines flight 1628[76], a Boeing 747-200F cargo aircraft, was en route from Paris to a layover in Anchorage, Alaska on its way to Tokyo. Pilot Kenju Terauchi, reported a seemingly inexplicable encounter in the skies as they approached the airport in Anchorage. He described a massive space craft, although no radars were able to confirm this. Over the next few months, further similar encounters were reported in the same area.

The Santa Catalina Channel[77] separates mainland California from the island of Catalina. This particular stretch of water is as deep as Mount Everest is high and UAPs have been seen both entering the water and emerging from its murky depths. But on June 14th 1992 CE, there were reportedly hundreds of UAP's. Independent witnesses claim that there were in excess of two hundred UAPs that emerged from the water, hovering momentarily before accelerating off into the sky at blistering speeds, all the while in complete silence. There were many witnesses, some of which were as far away as Malibu and many filed reports with their respective local police forces,

The UK's tests of their trident missiles in 2016 CE and 2025 CE which were conducted at a range in the USA, both failed soon after launch, leaving their test in 2012 CE as their most recent successful launch.

On 12th February 2023 CE, an octagonal object with strings hanging from it was detected over northern Montana, Wisconsin, and the Upper Peninsula of Michigan at 20,000 feet (6,096 m).

Airspace was temporarily closed in the Lake Huron area, where the object was shot down by the US Air Force and National Guard, falling into Canadian waters. This is reported to have caused a stand-off between USA and Canadian troops, over ownership but, as the USA got there first, possession proved to be nine tenths of the law. It was probably a Chinese spy balloon.

It is reported that many USA astronauts have seen UAPs following their capsules[78] [79], but have been told not to talk about them. They can switch to a secure radio channel to report them, although some have been heard to refer to "bogies" outside.

### iii. Canada

As Canada has vast areas off wilderness, the level of UAP reporting is comparatively low, with most emanating from near the USA border, where the main centres of population are.

Labrador and Newfoundland account for a number of sightings, including from military and commercial aircraft. The most significant occurred on 1st February 1967 CE at Shag Harbour[80]. An aircraft was seen to crash into the harbour, and local fishing boats went out to look for survivors or wreckage. They could find nothing but an orange scum. Police, who had also observed the crash, then confirmed that there was no aircraft missing, and it was realised that they had a crashed UAP on their hands.

The next day, naval divers were sent down, and it was claimed that they had found nothing. It was realised that the UAP still had underwater manoeuvrability, and that it had crept out of the harbour and into an adjoining inlet, where it was discovered by the Canadian Navy. However, it was not alone. A second UAP had arrived, and the crews were busy repairing the first. The naval vessels did not interfere, and eventually moved away in response to the detection of a USSR submarine many miles away. The two UAPs later rose out of the water and flew away to the north.

It is possibly not associated with the incident, but a lighthouse keeper nearby discovered a metallic object on the foreshore a few days later. He was not willing to say what was inside it, but he was ordered to hand it over to an USA officer who flew in specially to collect it.

### iv. USSR/Russia

The main sources of reports of UAPs and Others are official government papers which were obtained after the fall of communism in 1989 CE. The information here differs from that available from the USA in that, for long periods during the rule of communism and, after its fall, UAPs were officially recognised and the public and military were encouraged to report them. There was also an official committee to assess reports.

Russia has had its fair share of UAP sightings over the centuries, both before and after the Tunguska incident. It also had multiple sightings of Foo Fighters[81], in the skies and over the principal battlefields.

For many years, it was USSR policy to sent fighters to attack UAPs. However, they had already lost several aircraft before, in the 1960s CE, there was a full battle with UAPs[82] near their border with Iran and Afghanistan, where the USSR lost six fighters and their 12 aircrew. All fighter engagements with UAPs were banned in 1965 CE.

There were multiple sightings over the Kola Peninsular[83] in the early 1980s CE, and there were reports of UAP crashes.

In October 4th 1982 CE, UAPs appeared over a USSR ICBM base in Usove[84] in the Ukraine, performed astonishing manoeuvres in front of stunned eyewitnesses and then somehow took control of the launch system. The missiles which were aimed at the US suddenly started to initiate their launch procedures. Launch control codes were somehow entered, and the base was unable to stop what could have initiated World War 3. Then the UAPs disappeared just as suddenly, and the launch - control system shut down. This may have been the Greys giving a message to the USSR.

In 1983 CE, in Ordzenikidze[85], east of the Black Sea, a cone-shaped UAP was shot down by ground-based weapons. The wreck was quickly spirited away by the Russians.

At Dalnegorsk[86] in East USSR, on 29th April 1986 CE, an UAP was reported to have crashed into a hill (known as Hill 611). There were multiple UAP sightings over that hill thereafter, and the crash-site was somehow dangerously active for 3 years, with adverse effects to humans. Materials recovered were described as complex and not yet available on earth.

On April 26th 1986 CE, the Chernobyl[87] disaster occurred. Two technicians, who were dashing to the site after the first alarm, had to turn back when they measured the very high radiation. As they retreated, they saw a spherical UAP appear, which fired two red beams at the reactor for about 3 minutes, before departing. The technicians reported that the radiation levels fell from 3,000 milliroentgens/hour, to 800 milliroentgens/hour in that 3 minute period.

Although there had been occasional reports of UAPs buzzing USSR nuclear facilities before this event, their frequency increased significantly afterwards.

Russian Cosmonauts reported that, in May 1981 CE, the space-station Salyut-6[88] was approached by an unidentified space-ship to within 300 feet, and they could see 3 brown-skinned beings with slanted blue eyes. They communicated by holding up pictures to the windows, but didn't really get anywhere. The "Others" came out of their ship without any form of protective suit or breathing apparatus. It is thought therefore that these might have been some form of AI. The whole episode was filmed, but this evidence is not available.

Cosmonauts have also reported seeing angel-like entities[89] made of plasma, outside their space-station. At first, it was thought that this was caused by isolation or claustrophobia, but their replacement crew reported exactly the same phenomenon.

In 1989 CE, Russia sent two probes to the Mars moon, Phobos[90]. Unfortunately one never made it, and the second was lost soon after arrival, but not before it had photographed an UAP which was a cylinder 15 miles long.

UAP sightings have continued and, since the outbreak of the Russian-Ukrainian war, there are regular sightings over the battlefields.

v. United Kingdom

UAP sightings were not recorded at a national level in the UK until the 1950s CE[91], when 13 were recorded in that decade, including one at RAF Bentwaters, later to become famous as the site of the Rendlesham Forest Incident of 1980 CE.

On 4th February 1977 CE, it is claimed that a silver UAP landed on a field at the back of Broadhaven Primary School[92] in Wales, where it was seen by 14 pupils. The next day, their head teacher asked these to draw what they had seen, and the drawings were all remarkably similar. The same thing occurred on 16th February in an Anglesey Primary School, where 9 pupils drew very similar versions of what they had seen. However, this second sighting is considered less reliable, because it could so easily be a copycat report.

On 9th November 1978 CE a forestry worker claimed that an UAP had attempted to abduct him[93]. The police investigated and confirmed that the marks on his body, his clothing and the site where it occurred, all fitted in with his story, but could do no more. This occurred in what is now called the Falkirk Triangle[94], because of the large number of sightings recorded there. It is claimed that there are about 300 sightings there each year. This begs the question why? There may be an "Other" base nearby, and it is not far from Faslane, which is the home base of the UK nuclear armed submarines.

On 26th December 1980 CE, Just after midnight, eyewitnesses and radar screens at RAF Bentwaters, a USA air base reputed to store nuclear weapons, followed an unidentified object as it vanished into Rendlesham Forest[95]. Soldiers dispatched to the landing - site encountered a small luminous triangular - shaped craft, ten feet across and eight feet high, with three legs. The UAP then retracted the legs and easily manoeuvred its way out through the trees. The soldiers chased it into a field, where it abruptly shot upward, shining brilliant lights down on them. At that moment the witnesses lost consciousness. When they came to, they were back in the forest. Other troops sent to rescue them found tripod landing marks where the object apparently had rested.

The following evening, after observers reported strange lights, the deputy base commander, Lt. Col. Charles Halt, led a larger party into Rendlesham Forest. There, Halt measured abnormal amounts of radiation at the original landing site. Another, smaller group, off on a separate trek through the forest, spotted a dancing red light inside an eerily pulsating "fog." They alerted Halt's group, who suddenly saw the light heading toward them, spewing forth a rainbow waterfall of colours. Meanwhile, the second group now watched a glowing domed object in which they could see the shadows of figures moving about. During the next hour both groups observed these and other darting lights.

There were large-scale investigations into the incident by both the USA and UK, with claims of USA agents using drugs to interrogate personnel. It has been claimed that the whole thing was a psi-ops test, but it seems unlikely that the USA would try this on the guards to an important military facility, risking their sanity and hence the security of their weapons, when they could do the same thing in home territory.

In 1997 CE, the UK government received over 450 UAP sighting reports[96]. The UK public had clearly found someone to tell. This compares with only 9 in the whole 1970s CE decade. Of these sightings, there were about 30 large triangular craft, and the remainder were mainly bright lights or flying saucers, with a few cigar-shaped craft.

On 20th July 2008, a police helicopter had to take evasive action to avoid a UAP flying over the Bristol Channel[97], near the base at RAF St Athens. There were also a number of sightings reported over South Wales that day.

On 22nd February 2016 CE, residents of Pentyrch, a village in Wales, NW of Cardiff, reported a large pyramidal UAP[98], which hovered at a low level and discharged two smaller objects. Over the previous week, there had been reports of RAF Sentry aircraft regularly sweeping the area but now, a whole fleet of UK military aircraft appeared, and the UAP moved away. Two large explosions were reported. A Freedom of Information request for more details was refused on a variety of grounds. It is apparent that this UAP had been expected, and wasn't welcome.

Over the period 2020 CE – 2024 CE, UAP researchers gathered information from the police using Freedom of Information[99] enquiries and found that there were about 500 claimed sightings each year. The problem clearly has not gone away.

vi. Europe

In France, on October 16th 1954 CE, a Dr Robert[100] saw 4 UAPs, one of which landed in front of his car, and a small "Other" came out. All went dark for some time until the UAP flew off.

On October 27th 1954 CE, an UAP passed over a stadium soccer match in Florence[101], Italy. It was witnessed by over 11,000 spectators as well as people outside the stadium.

In October 1958 CE, a motorcyclist in Catalonia[102], Spain, stopped to assist a crashed plane. Instead it turned out to be a landed UAP with two small "Others" outside collecting samples from the local terrain, with one more inside. The witness watched for 15 minutes until they flew off.

In Belgium[103] on September 22nd 1965 CE, several witnesses, including Antwerp Airport employees saw a glowing sphere-shaped UAP flying at an estimated speed of 3,000 kph.

Over Donaghadee in Northern Ireland over a dozen people sighted several egg-shaped UAPs travelling at a phenomenal speed on 4th November 1975 CE.

From November 1989 CE to March 1990 CE there was a wave of triangular UAPs over Belgium. Many of the reports related a large object flying at low altitude. Some reports also stated that the craft was of a flat, triangular shape, with lights underneath.

The Belgian Wave peaked with the events of the night of March 30th, 1990. On that night, one unknown object was tracked on radar, and two Belgian Air Force F-16s were sent to investigate, with neither pilot reporting seeing the object. No reports were received from the public on that date but, over the next 2 weeks, reports from 143 people who claimed to have witnessed the object were received. Over the ensuing months, many others claimed to have witnessed these events as well. Following the incident, the Belgian Air Force released a report detailing the events of that night.

On 2nd September 1990 CE in Greece, there were reports of a UFO crash near the village of Megaplatanos[104] close to Atalanti. Around 3 am, shepherds and a small group of villagers observed 6 UFOs approaching from the north. One of the UFOs looked unstable. Bizarre lights emitted from the ships which did not make any noise. The troubled UFO lost altitude and crashed 500 meters or 1/3 of a mile from them. They did not hear any noise although the wooded area caught fire. The remaining UFOs stopped over the accident while two landed near the destroyed UFO. The fire was instantly put out and during the entire night, there was unusual traffic from the ground to the hovering UFOs, light spots went up and down collecting debris until sunrise. The UFOs disappeared right before sunrise. The entire village saw the event. The ground featured an oval-shaped burn with a cut pine tree in the centre. There were also very small metallic pieces and wires around the crash site. The Hellenic Air Force arrived hours later and informed the village that it may have been a Soviet satellite that crashed or a plane. The Hellenic Air Force took some of the pieces.

On November 5th 1990 CE, a triangular UAP[105] was spotted by air traffic controllers over Paris, Bischwilleer and Nantes. The UAP was also sighted by three aircraft crews in Italy.

In 2009 CE on 9th January On 1 AM, Adam Maksymów, of the village Jarnołtówek[106] near Prudnik in Poland, went outside to charge his car battery. He was interrupted by a noise which he likened to rockets blasting off. Then he heard a buzzing sound which he compared to that of a swarm of bees. He saw a blinding light, and a huge saucer with a triangular glowing blue beam on its underbelly rose above the ground and took off into the night sky at an impossible speed. Other people of Jarnołtówek also reported seeing the object.

I have only listed here a tiny fraction of the total list of UAP sightings across Europe. Overall there are thousands.

vii. Central America and the Caribbean

To the west of the main island of Cuba lie the waters of Guantanamo Bay[107]. Apart from its more notorious current reputation, there were also sightings of UAPs there in the 1960s.

Puerto Rico[108] is described as a massive UAP hot-spot, with the possibility of an UAP base beneath the lake Laguna Cartagena[109] and a second, underwater, to the north in the Milwaukee Deep. There have been multiple reports of UAP sightings, a child-abduction and sightings of a number of Chupacabras[110], or Goatsuckers[111].

Could it be that these Chupacabras are the actual "Others" here? They don't appear to harm humans, and it may be that they have a need to ingest blood to survive. This could be part of the reason that there are world-wide reports of cattle, sheep and horses being found drained of their blood. Are we here on earth helping to keep alive a species that needs blood to eat, but has the conscience not to attack intelligent species?

There have been many sightings of UAPs[112] apparently flying into Popocatepetl near Mexico City. This is an active volcano, but it is not the only one associated with UAPs. They could be openings to Portals or Gateways, bases for fire-loving ETs, or homes of ITs.

On August 25th 1974, north of Presidio[113], it is claimed that an UAP travelling at 2,000 miles per hour collided with a small plane. The flaming wreckage of both aircraft fell to the ground and the small plane was pretty well destroyed in the crash, but the UFO was intact, When the Mexican Army arrived, they found pieces of the plane and a silver disc approximately 17 feet across and six feet tall. Twenty-four Mexican soldiers surrounded the craft, amazed at their find, loaded it onto a flatbed truck and headed back to their base at Ojinaga. An hour later, all 24 soldiers were dead. The truck veered off the road and rolled to a stop. The U.S. was monitoring all this, listening in on radio communication between the soldiers on the ground and their headquarters. They flew a reconnaissance plane over the site of the truck and bodies. A flight investigation tcam was assembled and four helicopters took them from Fort Bliss in El Paso to an area about 20 miles south of Candelaria, Texas. The U.S. team, wearing biohazard gear, took it away. They then set off explosives to cleanse the area of contamination. The UFO was taken to a secret government base in the Davis Mountains, and then transferred to another military compound, probably Wright Patterson Air Force Base near Dayton, Ohio.

On July 11th 1991 CE, there was a total eclipse of the sun over Mexico, and thousands reported seeing UAPs[114]. There were many films of the event. Since then, there have been thousands of sightings throughout Mexico. There is no clear pattern, although the majority have occurred in and around Mexico City. The witnesses range from pilots to doctors to bus drivers and even school children.

On 5th March 2004 CE, a Mexican military aircraft[115] filmed a line of 11 UAPs flying near it, using a state-of-the-art infra - red camera. At one stage, the objects surrounded the aircraft. The Mexican government confirmed that it had released the video.

## viii. South America

There are claimed to have been sightings of large spacecraft hovering over generating plants and reservoirs there, causing power blackouts on a regular basis. Some tribes complain that they are stealing their water.

In Argentina, on 10th May 1950, it is claimed that a 32 feet diameter UAP crashed[116], killing the three "Other" occupants who were about 4 feet tall. The witness was an aeronautical engineer, who entered the craft, then left to get help. When they returned, there were two further UAPs overhead, and the crashed one had been reduced to a pile of ashes. Afterwards, he suffered from a fever for some weeks, and had blisters on the parts of his face which were not protected by the sun-glasses he was wearing.

In Venezuela[117], on 28th November 1954 CE, it is claimed that two hairy dwarf-like creatures and an 8-10 feet luminous sphere stood in the way of two truck drivers. One driver attempted to capture one of the dwarves, and lost the fight, fortunately with no major injury. The sphere flew off. On 10th December[118] there was a further sighting claimed, where dwarves again attempted to capture a young man, who managed to hit one with his unloaded shotgun, which shattered. The sphere flew off again. A similar incident is reported for the 16th December.

Back in Argentina[119] on 20th August 1957 CE, an Air Force guard claimed to have been approached by a disc-shaped UAP, from which came a voice claiming they were interplanetary beings who were setting up a base in Salta, to liaise with humans.

That year there were numerous reports of UAPs in Brazil, including one reported by Professor Guimares[120] who was approached by an UAP, from which two tall beings descended. He was taken on a short flight.

On 5th July 1965 CE and thereafter, in the Valle of Lorestani, in Argentina, there have been repeated sightings of UAPs about 10 – 15 metres in diameter, seen by dozens, including the head of the Argentine's UAP investigations, and the frequency has led to the suspicion that there is a base in the area.

In Peru in 1968 CE, there were many UAP sightings, leading to suggestions that they were using areas around Lake Yanacocha, Lake Titicaca[121], and various other lakes in the Cordillera Blanca.

On January 13th 1996 CE, the Brazilian Air Force is alleged to have shot down an UAP which crashed six miles from Varginha, a medium-sized town in south-eastern Brazil[122]. Seven days later, two sisters aged 14 and 16, and a 21-year-old friend spotted a tiny, frightened alien with big red eyes, crouching by a wall. They ran screaming back to their mother. The Brazilian police and military captured at least two aliens, one of which scratched an officer, infecting and ultimately killing him, before dying along with its extraterrestrial comrades. The US Air Force confiscated the alien bodies and took them to an unknown location. A vast cover-up by the Brazilian military, enforced with death threats, lasted for 26 years.

The above examples are just a few of the many alleged sightings recorded in South America in this period. With the growth of air travel on the continent, there has also been a commensurate growth of claimed UAP sightings in the air, including the sighting on 5th October 1996 in Brazil[123], of a massive UAP, cylindrical in shape, and 300 – 400 feet long.

In Argentina on the night of 5th September 2023 CE, in the Espora Air Base, near Bahía Blanca[124], four UFOs flew over the base and were seen by military personnel, who responded with gunfire. The objects were black and triangular in shape. One object fired with a type of laser that injured two or three soldiers. The Base denied any incident and claimed that the videos and audios that circulated that night were edited to create fake news.

### ix. Japan and the Dragon's Triangle

The Dragon's Triangle (Devil's Triangle) is loosely defined as an area south of Tokyo, to the Philippines and the Marianas Islands.

However, in his book on the subject, Charles Berlitz[125] even looks at incidents in Tasmania, south of Australia and also in New Zealand! Great care needs to be exercised in using his findings.

The Dragon's Triangle has a reputation at least equal to the Bermuda Triangle, with similar fuzziness about what lies within its limits, and what is outside. Many of the incidents recorded there may not involve UAPs, as it has notorious waves and underwater volcanoes, but many believe that there is an underwater UAP base there.

On February 21st 1957 CE, the citizens of Yokohama City[126] watched as fleets of UAPs flew over them in V formations. The whole affair lasted seven minutes, and they were totally silent.

When sailing back to Tokyo from the south[127] on April 19th 1957 CE, a fishing boat encountered two metallic very silvery UAPs descending from the sky, which dived into the water.

On March 21st 1965 CE[128], an UAP followed a TOA airlines plane for 55 miles, causing some electronic systems to fail. The pilot was able to use his radio to tell traffic controllers what was happening and a second aircraft then claimed that the same thing was happening to it.

In 1975 CE on 23rd February[129], two seven year old boys watched a bright silvery UAP, shaped like a domed disc, land in a vineyard in Kofu. In it was a humanoid creature, dark skinned with pointed ears, and a wrinkly face with no other features apart from fangs. It spoke unintelligibly and the boys ran off. Their parents saw orange light in the vineyard as the UAP flew off.

When sailing in the Devil's triangle on 17th April 1981 CE, a freighter, the Taki Kyoto Maru[130] suddenly lurched as if it had been struck by something. A 50 feet diameter flying saucer rose out of the sea, and circled the ship for about 15 minutes. It then dived back into the sea, causing waves which almost capsized the ship. The captain then found that the ship's clocks had lost those 15 minutes.

Between 1946 CE and 1986 CE, there have been a total of about 25 large surface ship disappearances[131] in the Dragon's Triangle. In that same period a total of 13 soviet submarines were lost in the Japanese area, together with any nuclear weapons which they were carrying. However, only 2 of these submarines were in the Dragon's Triangle. The others were in the Sea of Japan or the South China Sea.

Close by this arbitrary triangle is the island of Pohnpei in Micronesia. This contains the remarkable ruins of Nan Madol,[132] and nearby are claimed to be two legendary submerged prehistoric cities known as Kahnihmw Namkhet and Kahnihmweiso. There is little archaeology to support this claim, but it could be that this was the location of an "Other" base.

In the mountainous region of Fukushima Prefecture[133] in Japan, residents report frequent sightings of luminous objects near Senganmori Mountain, and of "Others" with wings and hawk-like features. These Avians sound similar to Anzu of ancient Sumeria or Horus of Egypt.

### x. Australia & New Zealand

On January 19, 1966, a calm sunny day, a banana farmer named George Pedley was driving a tractor near Horseshoe Lagoon[134] near Tully, in tropical far north Queensland, Australia. When he was about 25 yards from the lagoon, he heard a loud hissing sound above the noise of the tractor. Suddenly, an object rose out of the swamp. When he glanced at it, it was already 30 feet above the ground, and at about tree-top level. It was a large, grey, saucer-shaped object, convex on the top and bottom and measured some 25 feet across and 9 feet high. It rose another 30 feet, spinning very fast, then it made a shallow dive and took off with tremendous speed. Climbing at an angle of 45 degrees it disappeared within seconds in a south-westerly direction.

When he came to the spot from which the object had risen, there in the lagoon was a large circular area that was clear of reeds and in which the water was rotating slowly. A few hours later, at about noon, he returned to the lagoon for a second look. The scene had changed, because now the circular area was covered by a floating mass of green reeds that were distributed in a clockwise radial pattern. The circular mass of reeds was about 30 feet in diameter.

Oddly, the outside edges of the mass of reeds angled down, similar to the shape of a saucer placed face down. He reported his experience to the Tully police that evening, and they in turn reported it to the RAAF after making a trip to the site the next day, January 20$^{th}$. During the course of the investigations, as many as five other "nests", all smaller than the original, were discovered. In some of these, the reeds were rotated in a counterclockwise direction and a couple of them showed signs of burning in the center of the nest.

On 6th April 1966 CE, students and a teacher from Westall High School[135] (now Westall Secondary College) reported seeing a flying object. It was described as round with a domed top, and white, grey, or silver in colour. According to the students, the object descended behind a row of trees and into the Grange, an open area south of the school. Some accounts describe the object as being pursued by five unidentified aircraft. Shaun Matthews was on vacation at the Grange and reported seeing an object with a slight purple hue and about twice the size of a family car.

Some witnesses reported seeing the object take off after landing, and some reported seeing it hover rather than land. When students walked to the Grange after the sighting, some reported a landing site, but the details varied between reports. Students variously described a circle of grass as burnt, "boiled" or pressed down. One student interviewed by a local newspaper described a vague circular area flattened by the wind. Students also reported varying numbers of circles from one to three.

The Kaikōura lights[136] is a name given by the New Zealand media to a series of UAP sightings that occurred fist on 21st December 1978 CE, over the skies above the Kaikōura mountain ranges in the northeast of New Zealand's South Island. The first sightings were made when the crew of a Safe Air Ltd cargo aircraft began observing a series of strange lights around them, which tracked along with their aircraft for several minutes before disappearing and then reappearing elsewhere. The UAP was very large and had five white flashing lights that were visible on the craft. The pilots described some of the lights to be the size of a house and others small but flashing brilliantly. These objects appeared on the air traffic controller radar in Wellington and also on the aircraft's on-board radar.

On 30 December 1978 CE, a television crew from Australia recorded background film for a network show of interviews about the sightings. For many minutes at a time on the flight to Christchurch, unidentified lights were observed by five people on the flight deck, were tracked by Wellington Air Traffic Controllers, and filmed in colour by the television crew. One object reportedly followed the aircraft almost until landing. The cargo plane then took off again with the television crew still on board, heading for Blenheim. When the aircraft reached about 2000 feet, it encountered what appeared to be a large lighted orb which fell into station off the wing tip and tracked along with the cargo aircraft for almost quarter of an hour, while being filmed, watched, tracked on the aircraft radar and described on a tape recording made by the TV film crew.

They have appeared intermittently since the initial December 1978 CE sightings, with the most recent sighting being reported during 2015 CE.

A Darlington[137], Perth, man says he's captured pictures of hundreds of UAPs from the veranda of his home. It began when he was taking photos of clouds to test out a new camera and he noticed a "smudge" that, when enlarged and enhanced, "had some structure to it, suggesting it could be some sort of craft in the sky". He says that, since then, he has identified a dozen different UAPs including round, square and saucer-shaped craft, posting the photos to his website wispyclouds.net for extra-terrestrial buffs and sceptics to ponder.

He takes about 30 shots at a time. In 10-15 minutes he'll take 300 to 400 images. Then he connects his camera to the computer, zooms in and enhances any little thing he notes on the images. He gets UAPs in anywhere from 2 per cent to 20 per cent of shots. Some of them appear to have transparent canopies and in some shots it looks like there could be occupants inside. There is always doubt, but UAP stands for unidentified anomalous phenomenon and as far as he's concerned these aren't identified. It's possible some are man-made, but he doesn't think they all are.

Pine Gap[138], near Alice Springs, was originally set up as a USA eavesdropping station to listen in to China. However, it has been much enlarged with great warrens excavated beneath it, and it is now thought that it is acting as a liaison base for "Others". There have been a number of reported phenomena.

One of the strangest incidents occurred in 1973 CE[139]. A cartographer, working for the Australian government, was parked near to the Pine Gap base. It was late – just after midnight. Out of nowhere, a strange but intense "vertical shaft of blue light" came from the confines of the base. Giving in to his curiosity he drove his vehicle closer. He was shocked to see a disc-shaped craft hovering around a thousand feet in the air over the base. He raised his binoculars to his eyes. As he did, another blast of cold, blue light emerged. It was coming from the center of the craft and heading down to the domes of the base. After several moments it went off. Then, a similar laser-like light extended from somewhere on the ground to the disc above. This went on for over thirty minutes before the disc began to spin rapidly. It then shot off into the night sky at great speed.

At some time during 1984 CE, an intense gold pillar of light[140] shot upwards from the middle of Pine Gap base grounds. It was several meters wide and appeared to be solid as it stretched upward into the night sky. Above them, and around the point of the light, a strange cloud appeared to be forming. As the light "pulsed" the cloud appeared to grow. Just as suddenly as it began, the light shut off. Before the five witnesses could get their bearings, however, the light appeared again, only this time coming down from above and hitting the grounds of the base. It was at this point that the witnesses saw for the first time five strange objects above the base. Four of these were in a "diamond formation", while behind them was a cylindrical object. The first four objects moved into north-east-south-west positions, while the cylinder-shaped object positioned itself in the middle of them. Then an intense solid light, this time blue, hit the ground from the objects above. Then strange cloud formations also appeared high up the beam. This continued for several minutes before the lights shut down and the five apparently alien crafts slowly rose upwards before vanishing in a flash. Although none of the witnesses knew what they had just seen, all agreed that the rumors of established contact with an extra-terrestrial race were apparently true.

At around 4:30 am in the early hours of 22nd December 1989 CE, three men, while returning from an all-night hunting trip[141] near to the Pine Gap facility, suddenly noticed activity in the grounds of the base. A camouflaged door suddenly opened before them, revealing lights and movement behind it. From inside this hidden enclosure, a metallic, grey disc emerged. No sound accompanied its movements, and aside from their own breathing, all around them was silent. Suddenly, but still with little sound, the disc shot off at an amazing pace, certainly beyond anything any of the witnesses had ever seen. The door then calmly shut, hiding its presence once more in the process.

## xi. Worldwide

UAP appearances are in no way confined to the various areas I have detailed above. They are a worldwide phenomenon as are portals and gateways. Cattle mutilations occur all over the world, and not just of cattle. Both sheep and horses are subject to attacks as well. It would appear that ETs did not really need permission to take cattle in the USA. They would have done it anyway.

China has a long history of their larger mountains being considered "Other" bases, with UAPs being sighted in their vicinity. It has been suggested that the Chinese concept of a dragon comes from sightings of flame-spouting UAPs being miss-interpreted. China's first Emperor Qui Shi Huang is thought to have either been "Other" or closely assisted by "Others". There have been recent photographs of apparent UAPs over major airports[142], resulting in their temporary closure.

In Africa, the principle items of interest are the pyramids in Egypt and the Sphinx. However, there are many reports of lights in the sky and of circular UAPs of various sizes and a couple of reports of large triangular UAPs. There is at least one claim of an UAP landing 1989 CE, in the Kalahari Desert[143], where cattle herders described strange very tall humanoid "Others" exiting. They wisely fled. Nearby the same year, it is claimed that a UAP crashed[144], with a South Africa military cover up.

Worldwide, wherever there is still a tribal sense of identity, as in Africa, part of their oral tradition is that "Others" came to teach them. In places, rock art preserves a record of this, as well as carvings.

### xii. Summary

Having looked at the vast number of sightings that have been recorded around the world, I am left in no doubt that UAPs are real. Many of the sightings are of lights in the sky, rather than of identifiable spacecraft, but these lights are simply "Other" reconnaissance drones, of vastly superior abilities to those we can make for ourselves. There are enough sightings of larger UAPs to justify my conclusion. Beyond that, the attitude of governments towards them, makes it clear that they know all about them, and are either content to leave them to their own devices, or are too frightened of them to do anything.

## d) WHEN THEY MAKE CONTACT

### i. Contact with the Authorities

Now that I have come to the conclusion that UAPs are real, I also have to conclude that they have been in regular contact with many governments since at least World War II. It is not my place to justify their reasons for withholding this information.

ii. Contact with Real People

In June 1920 CE, Albert Coe[145] was canoeing in Ontario when he rescued a stranger from a ravine. He had to carry him to his craft which turned out to be a 20 feet diameter flying saucer. He watched it take off, and saw him afterwards 10-12 times a year until the 1970s CE. The “Other”, named Zret, had crash-landed on Mars with 3700 “Others,” eventually colonising Earth in Atlantis, the Cuzco Valley in the Andes, Lemuria (near the Marshall Islands), Tibet and Lebanon. Their main current concern was Earth’s potential to develop atomic weapons. He looked fully human.

In May 1940 CE, a tall humanoid with white hair descended from a flying saucer at Helena[146], Montana in the US, and asked for assistance in getting water from a nearby stream. They said that they were there to monitor the development of humans, but were not allowed to interfere. They were probably Nordics.

On 4th July 1949 CE, Daniel Fry[147], a rocket test technician at White Sands in the USA, saw an UAP land and, as he approached, a voice warned him not to touch as it was dangerous. The voice turned out to be that of an “Other” who called himself Alan, although he never saw him. He said his purpose was to see how adaptable humans were to new ideas. Daniel was given basic descriptions of how the UAP’s mechanism worked. He was also told how they had developed a previous earth civilisation on Atlantis, which had been destroyed by war, and the survivors had fled to Tibet.

On 11th April 1952 CE, near Nimes in France, Rose C[148] was woken by her dogs and, going outside, met 4 "Others", three being very tall. The leader explained that they had established earth thousands of years ago, as a penal colony. About 6400 BCE, there was a man-made cataclysm on earth, destroying their civilisation. Their job now was to take vegetation and soil samples following our use of atomic bombs.

On 20th November 1952 CE, George Adamski[149] claimed to have met the occupant of a cigar-shaped UAP in the Californian Desert. Called Orthon, he was long haired and about 5' 6" tall, and said he was concerned about radiation from nuclear bombs. Adamski was invited to board the UAP but declined and Orthon flew off. Adamski then claimed that Orthon returned, and started trading on his sighting, and claimed that he had gone aboard the UAP. Over time, his claims became more and more extravagant, until eventually his most ardent supporters gave up on him. His secretary left in the 1960s CE, citing his oversize ego as the reason why the "Other" gave up on him.

When he was 10 years old, in 1932 CE, Howard Menger[150] first encountered an "Other", a beautiful tall young woman with golden hair – probably a Nordic. He had further encounters with male and female Nordics regularly for the next 65 years. They told him of the coming of atomic bombs. They asked him to acquire earthly clothing and fresh food (although they weren't too keen on it). They took him on trips to the moon, visiting their base there. He was taken to their earth base in the Blue Mountains of Virginia, where it is believed they were mining beryllium, zirconium and titanium. Although he enjoyed the notoriety when he first spoke about his experiences, he quickly came to regret it.

### iii. What to believe?

The stories in the section above could all be fantasies dreamt up by attention seekers. The "Others" described here all seem benign, and it may be that you don't hear of malignant "Others", because no contactee survives them. Nevertheless, these are all described as having concerns about our nuclear weapons, and our polluting our world: and we have no evidence that our governments have the same worries.

# Chapter 3

## SO WHAT ARE THESE "OTHERS" ?

### a) INTRODUCTION

It is probably far more likely that UAPs are "Other" in origin, although this does not mean that they are necessarily Extra Terrestrial. However, this still gives far too many choices:

- They are caused by gods or magicians
- They are from our future
- They are from an earlier human civilisation in hiding
- Apes weren't the only earthly creatures to evolve
- They come from parallel universes
- They are products of an "Other" culture
- They are human products of reverse engineering of "Other" technology

### b) THEY ARE CAUSED BY GODS OR MAGICIANS

Most folklore talks about Gods. They generally descend to earth and do amazing things, often to the benefit of the local humans.

Some stories imply that these Gods can get very angry, and even fight each other.

The distinction between a God and a Magician is a matter of personal belief. They can both do apparently impossible things. But, as Arthur C Clarke said "Any sufficiently advanced technology is indistinguishable from magic". I have always wondered why, if a being is as far technically advanced from me as I am from an ant, they should want us to build churches and worship them. I would be horrified if I found that ants were worshipping me.

So for the moment, as I am not clear what makes a being a God, I am simply going to define Gods as belonging to a superior civilisation.

## c) THEY ARE FROM OUR FUTURE

There is little evidence to support this hypothesis, apart from some photographs of anachronisms, which could be simple miss-interpretations. The classic possibility of time paradoxes suggests that it would be unwise for our descendents to try to change us lest, in doing so, they change themselves. There is the possibility that they could be sight-seers from the future. There are unconfirmed reports of UAPs watching the first moon-landing. As this was one of main scientific events of the twentieth century, I would love to have had the chance to watch!

## d) THEY ARE FROM AN EARLIER HUMAN CIVILISATION, IN HIDING

Given that the time interval between the extinction of the dinosaurs, and the generally accepted evolution of the modern human some 20 thousand years ago, is some 66 million years, it is not beyond the realm of possibility that a fraction of modern humans actually started their final stage of evolution say 30 thousand years ago. This would give that fraction a 10 thousand year lead on most humans in developing civilisation. They would today appear very like the rest of humanity, and could co-mingle without detection.

In the extra 10,000 years of their civilisation's development, they could have fully explored the Solar System, and could be living out there, leaving the earth to us apart from the occasional hidden base. They could even have been the "Gods" who came back to educate early mankind.

Equally, as described in the previous chapter, they could have been wiped out at some stage. Several "Others" have told contactees this. What is missing from archaeological records is evidence of their using resources beyond stone. If they had developed a society similar to our present one, there would be mineral mines for us to find. They would have to have gone down a different technological route which did not involve the use of metals to any great extent.

## e) APES WEREN'T THE ONLY EARTHLY CREATURES TO EVOLVE

Whilst humans were evolving in the 66 million years since the extinction of the dinosaurs, all the other creatures on earth were evolving too. To have avoided the dinosaur extinction event, they would probably have to have lived deep underground or deep underwater, may even have developed more than primitive mammals before the extinction event, and could have got a running start towards civilisation. They could be much more advanced than humans.

If they lived in very deep caves, or underwater beyond the continental shelves, it is quite understandable that we have not yet found any archaeological record of their existence.

These are what are termed Intra-Terrestrials (ITs).

## f) THEY ARE PARALLEL UNIVERSE ENTITIES (PUEs)

Physicists, working in the field of String Theory, seem happy to contend with multiple universes, perhaps even an infinite number of them. It may be possible to move from one universe to another via Portals.

One proposal resulting from string theory is that, whenever for example there are two possible answers to a decision, the universe splits into two new universes, one for each possible outcome. As this is happening everywhere, all the time, we end up with an infinite number of possible universes. This poses a problem when we want to move between universes. In many cases, it would not be possible to tell one universe from many others, until the results of all the possible outcomes have worked through enough to show a significant difference. This might take years. Even if one developed a way to move from one universe to another by a Portal, it might be very difficult to find one where things were drastically different. How long ago would the change have to have happened before a universe had a significantly different technology from ours, let alone to appear physically different?

If we are the subject of incursions from a parallel universe, this would suggest that there might be far fewer such universes. The inhabitants of one of these would need to be able to create Portals, through which they could visit us.

If Portals between parallel universes could open of their own accord, this might account for some of the exotic creatures which are claimed to exist, but do not appear to be in sufficient numbers to constitute a breeding population. These cryptids might include Mermaids, Skinwalkers, Unicorns, Werewolves, Mothmen, Chupacabras and many more. Of course, they may simply be entities here for a vacation! It is difficult to imagine secret bases for Reptoids in the swamp areas they appear to enjoy, but they are reputed to be intelligent, if aggressive.

I have deliberately excluded Bigfoot, (including Yeti, Sasquatch, Yowie, etc) on the grounds that there have been so many sightings that, even without any definitive proof of their existence, it is credible that there could be sufficient numbers present for a breeding population to exist.

Small Portals are sufficient to permit the passage of the small UAPs generally described as globes of light or small metallic objects, which seem to be reconnaissance devices to see "what are these humans up to now?" Native Americans have shown that they are capable of opening small portals using rhythmic chanting and drumming. It is claimed that this was filmed in the TV series "Beyond Skinwalker Ranch". In the Indian sub-continent, the use of chanting appears to serve the same purpose, with mantras claimed to be very powerful.

Larger Portals would be necessary to permit the passing of larger UAPs, which could be described as means of transport for entities of approximately human proportions. This might generally require more energy than could be achieved by drumming and chanting, but could still require some sonic trigger.

If some parallel universes could be considered to be near each other, it might be possible for events in one universe to have an effect on another. This might account for reports of PUE visitors being concerned about our use of nuclear weapons, and their being particularly interested in nuclear tests, nuclear stockpiles, nuclear-armed rockets, aircraft and ships, and nuclear power plants. Whilst any race in any universe or galaxy may well be altruistically concerned that we shouldn't wipe ourselves out, they may well be much more concerned if we might wipe them out too.

Certainly, if we find signs of crossing points into our world, we don't have to assume they are wormholes, or that our visitors are from outer space.

## g) THEY ARE PRODUCTS OF AN "OTHER" CULTURE

### i. General

Firstly we need to remember that there are many possible different types of "Other" culture. As explored in the previous section, there is the possibility of a large number of cultures in parallel dimensions (PUEs), which could reach us by Portal. Secondly, if long-distance travel within our own dimension is possible, whether or not by Wormhole or Gateway, there could be an almost infinite number of planets where Extra -Terrestrial (ET) cultures might be much more technologically advanced than us. Thirdly and often ignored by Ancient Astronaut Theorists, there is the possibility of a limited number of Intra-Terrestrial races (ITs) living on this planet either deep underground or deep underwater, and which could again be more advanced than us. They would certainly be concerned by our using nuclear weapons.

It is very concerning to hear claims that Occam's Razor can be applied to prove that "Other" cultures, which have been visiting us for millennia, must be ET. The simplest solution should not be the one which requires the most complexity.

### ii. Intra-Terrestrials (ITs)

It is quite likely that this possibility has not intruded greatly on earthly consciousness. After all, we should have seen them, or their bodies, or archaeological evidence, shouldn't we? This is not necessarily so.

Let's begin by considering things from a historical perspective. The dinosaurs were wiped out approximately 66 million years BCE both on the land and in the sea. At that time there were small mammals which survived, and from which it presumed that humans and other mammals evolved. Presumably these small mammals would have to have lived underground to have avoided the extinction events. The skull of Lucy, thought to be an early hominid, was dated to 3.2.million years BCE. Between these dates, the last ice-age started about 34 million years BCE and continued until relatively recently at about 11 thousand years BCE.

66 million years is the time it took for humans to evolve, but who is to say that other species – ITs - did not evolve in the same period, perhaps more rapidly than us? There are apparently no fossil records to suggest this. However we have not yet searched for fossil records in ultra deep cave systems or the depths of the oceans.

Native American Hopi folklore speaks of Ant People[151] living deep underground, possibly somewhere in the Rocky Mountains. African folklore speaks of gods coming from the water. Other folklores speak of holy mountains where their old gods lived. These dwelling places may, of course, be sites which were temporarily occupied by "Others" on their visits to earth. Many cultures describe visits from their gods, teaching them agriculture, mathematics, medicine and how to build. These gods could be ETs, ITs, a more advanced human civilisation, or entities from a parallel universe.

It would appear that there may have been human migration as part of the mitigation measures taken by the ETs or ITs to protect humankind from the great flood, resulting from the melting of ice at the end of the Ice Age.

Many cultures start by the sea. Their first towns would have taken advantage of fishing for food, rivers to provide fresh water, and higher coastal rainfall for irrigation. These towns would be under threat from a flood. Underwater towns and cities are now being discovered, on the continental shelves but below the present sea level – Dwarka[152] in India, Pavlopetri[153] in Greece, Atlit-Yam[154] in Israel, and Mahabalipuram[155] in India are such examples.

The underground city at Cappadocia in Turkey can accommodate 20,000 people. It is difficult to imagine this being excavated by humans back in 12000 BCE, but might have been constructed by ITs or ETs to shelter refugees from the floods. Some Native Americans are now desert dwelling, and their folklore describes their being taken underground for shelter by their gods. Perhaps these survivors were originally coastal dwellers and were moved to safety well inland or to higher ground.

### iii. Extra Terrestrials (ETs)

There is a fair amount of evidence to suggest that ETs have visited us in the past. There are drawings and carvings showing entities clad in what appear to be space helmets. However, whilst ITs would not normally need space helmets, it is possible that PUEs might. In fact, almost everything attributed to ETs could, with few exceptions, be attributed to advanced humans, ITs and/or PUEs.

At a primary level, the unique capability required of ETs is that they can travel vast distances in space by using a Wormhole, Gateway or some other technology. It is rightly argued that their civilisations could be billions of years more advanced than humans, and their technology might make this possible. However, the converse need not be true – the presence of visiting entities does not prove that wormholes or some such technology must exist.

It has been postulated that ET entities are altruistically motivated to develop the human species, teaching us many basics, such as farming, mathematics, medicine and astronomy. It is suggested that many of the large-scale building projects of the past were undertaken by, or with their assistance.

It is also hypothesised that their presence could have been motivated by the search for raw materials, particularly gold. The claimed genetic discontinuity in the development of humans could have been introduced by them to generate a workforce. It is interesting that practically every part of human society, worldwide, has always valued gold extremely highly. We also value it today because of the many ways we can use it in electronics. We have been steadily gathering all the gold we can, and storing it for the future. However, there have been persistent rumours that there is gold missing from Fort Knox, and an audit is proposed[156]. If gold is missing, will the USA ever admit it? Perhaps we have finally made our first payment to the ETs.

## h) THEY ARE PRODUCTS OF HUMAN REVERSE ENGINEERING OF "OTHER" TECHNOLOGY

During World War II, Germany was reportedly attempting to reverse-engineer a spaceship which crash-landed there in 1936 CE. There is no record of what happened to it after the war, but it is fairly certain that either the USSR or the USA spirited it away. Given the speed with which the USA started to advance technically after the war, they would seem the better bet.

Two subsequent crashes at Roswell (1947 CE) and Kingman (1953 CE) may have further helped, but the major benefit of these was that there was an "Other" survivor from these crashes, who assisted with the reverse-engineering project for many years.

Although there have been a fair number of whistle-blowers, including Bob Lazar, there has never been a public admission that they are reverse-engineering flying saucers. There have, however, been claims that a number of technical advances have been facilitated by this process, including fibre-optics, stealth technology, night vision and micro-processors.

It is possible that some of the UAP crashes since the Springfield crash may have been of USA prototypes. Scrapped USA jet-powered circular aircraft prototypes were reported in a scrap-yard at MacDill Air Force base in 1967 CE. It is also rumoured that the USA has a way of shooting down UAPs. One possibility is the use of Electro-Magnetic Pulses (EMPs), and some sort of beam weapon[157] has been witnessed firing from Area 52 which is another name for Dugway Proving Ground, a military installation in Utah.

Since then, it would appear that the USA has been developing a flying triangle, called the TR3B[158], which does not appear to be jet-powered, but may have been part-dirigible at first. This suggests that the reverse-engineering programme, and the assistance of the "Other" J Rod, has paid dividends. Some flight testing was done in the Hudson Valley in the 1980s CE, followed by deployments over Belgium in the 1990s CE as well as over England in 1993 CE. There were many reports of these vehicles apparently following patrol routes down to the east of the Rockies and once over Phoenix, Arizona in 1997 CE, viewed by thousands and known as the Phoenix Lights.

Given that at least 30 years have passed since flying triangles appear to have become operational, there could be, by now, a third generation of USA UAPs that we haven't seen yet.

The extremely aggressive way in which some element within the USA has attacked eyewitnesses, and UAP researchers, sending its thugs to threaten them and their families, suggests a grim determination to conceal their activities, offering virtual proof in the process. Of course, this may be in response to the agreement between Eisenhower and the Greys, dating back to 1954 CE.

This at least accounts for one group of UAVs.

# Chapter 4

## WHERE ARE THEY?

### a. WHERE ARE THEIR BASES?

#### i. Elsewhere

To state the obvious, it is inevitable that our Moon should be considered as a primary base, with claims that it is artificial and hollow. There are also claims that there are "Other" facilities on the back of the Moon. The space-going earthly nations will know the truth of this, but they still cling to their charade of denying everything, far beyond what might be considered reasonable.

It is also reasonable to assume that, if there are "Others" on Earth, they may well also have local bases elsewhere in our Solar System.

#### ii. Underground

One of the most popular proposed locations for an "Other" base is Antarctica. It is suggested that Nazi Germany and the USA found it in the 1930s-40s. It is claimed that Admiral Byrd, during Operation High Jump, was taken into an under-ice city to meet one of the "Other" leaders, who lectured him on the perils of the

Atom Bomb. Byrd then aborted his mission, dashed back to Washington to report and was ordered to keep silent. It is claimed that his personal diary has been found. There have since been many reports by scientific and military personnel stationed in Antarctica of strange events, sightings and encounters. It should be noted that Admiral Byrd did not comment on the appearance of the "Other" he met. This could be presumed to mean that the person he met was human in appearance, or he would have said something. Antarctica is possibly a base for the earlier earth-human civilisation. They are reported to have flying saucers, and possibly cigar-shaped UAPs like the one which is supposed to have crashed on a glacier on the island of South Georgia. This is a society to blame for some of our UAPs.

When "Others" first made contact with our ancestors, they were described as living in mountains such as Mount Kailash in Tibet, Mount Musiné in Italy, Mount Shasta[159] in California and Mount Denali[160] (formerly called Mount McKinley before reverting to its Athabascan name) in Alaska. There are reportedly many others but some, such as Mount Olympus, do not appear to be occupied any longer. Also UAPs are reported flying into volcanoes such as Popocatépetl in Mexico and Mount Shishaldin[161] in the Aleutian Islands.

Mount Kailash in Tibet is generally associated with a gigantic underground city, and there are present day reports of UAPs flying into it through doorways which appear and disappear. If it is occupied by "Others", there seems to be no evidence to identify what race they belong to, although it is possible that the blue-skinned "Others" from India are there now. However, the sightings are of flying saucer type UAPs, not of the Vimanas, their more conical-shaped UAPs. To live in such a city, they would need to have curbed their argumentative natures. It is interesting that the first representations of Buddha show him as blue-skinned like them, so perhaps they have.

Mount Musiné still gets visits from UAPs, although claimed sightings appear less in number. Mount Shasta in the USA is still a hot-spot of the unexplained. There are reported disappearances, sightings of UAPs, and reports of strangers in the caverns vanishing into the rock walls. These have been described as typical humans, so this may also be a base for the earlier earth-human civilisation. Mount Hayes in Alaska appears to be under investigation by the USA. It is claimed that the mountain contains a massive black triangular pyramid, which is generating so much power that it could supply the whole of Canada. It is not known which "Other" race produced this, but it is reminiscent of claims made about the great pyramid in Egypt.

The Falkirk Triangle in Scotland has so many claimed sightings every year, that it is a distinct possibility that there is an UAP underground base in the vicinity.

There is some indication that there is an underground base somewhere in Argentina, probably on the western slopes of the Andes.

Multiple sightings of the smaller illuminated globe type of UAP are reported world-wide, both accompanying larger UAPs and acting in small numbers. It has been hypothesised that these are generally used for reconnaissance although, ever since the World War II Foo Fighters, they are known to have an offensive capability - disrupting electronics just by their proximity. Many of these have been sighted in the vicinity of Skinwalker Ranch in Utah, which has been the subject of intensive investigation for years, and electronic failure is the norm. Native Indians suggest that the Ranch is the site of a portal, and that cryptids have been seen there. There is a spiral of rocks there which they say was constructed as an indicator of a portal. A spiral carved into a rock face is common worldwide and may be an indicator that portals are a global phenomenon. Native Indians suggest that these portals are to Parallel Universes, so it could be that PUEs are watching us closely, without participating actively.

In Dulce in the USA, there is reputed to be a massive underground base beneath Archuleta Mesa, housing up to 18,000 "Others", primarily Greys, in the lowest 3 of 8 levels. In the Dulce area there are reports of UAP sightings as well as cattle mutilations. There was reported in 1979 CE that there was a fire-fight in the base, probably started by a trigger-happy human, but no-one is sure of the outcome. This could be the base originally agreed in the reported deal with President Eisenhower.

There is deep suspicion that the joint CIA/Australia listening station in Pine Gap, south of Alice Springs, has expanded significantly and it is claimed that it may now be a joint USA/"Other" base.

### iii. Underwater

It is claimed that about half of UAP sightings are associated with water, in one way or another. Most of these are over land or coastline, and it is suggested that they use water in their propulsion system.

It is perhaps inevitable that there are far fewer UAPs sighted out at sea than there are on land. Of these, most are reported near to the coast by ordinary people, whilst those in the deep sea are mostly by mariners. Sightings by naval personnel tend to be censored.

In 1982 CE, USSR divers in Lake Baikal observed underwater "Others" and made the mistake of trying to capture one. All the divers were suddenly forced to the surface, with three dying from the bends. These "Others" were described as humanoid in appearance, but adapted for life underwater. It seems unlikely that a fully aquatic "Other" would need to evolve legs. They would have to have evolved from amphibians.

There are claims of underwater bases near Cuba, Puerto Rico, Malibu in the USA, and in Lake Fluxian in China. The Cuban and Chinese bases seem now to be uninhabited. The Malibu base could account for all the UAP sightings near Catalina Island and the USA Vandenberg rocket base. There have been no claims about what type of "Other" lives there, but they could be ETs or ITs.

In 1956 CE, an "Other" contactee claimed to have been told of three underwater bases, one on the Uruguayan coast 25 miles from Buenos Aires[162], and one 100 miles south of Rio de Janeiro[163]. It was also claimed that there was one somewhere in the Gulf of Mexico.

In June 1959 CE, it was claimed that a UAP emerged from the sea in Tierra del Fuego, Argentina. This was observed on two separate occasions.

In 1963 CE, it was claimed that an underwater UAP interrupted a US Navy exercise off the coast of Puerto Rico.

There are reports of an underwater base on the deep water off the Bahamas, which has power and communication links to an USA base, AUTEC, on the Andros Island[164] coast. This could be another example of "Others" working with the USA, like in Dulce in New Mexico, and Pine Gap in Australia.

There are possibly two UAP bases in the Japan area. One could be in the mountains in the very north, and one underwater in Micronesia in the south of the Dragon's Triangle

Given that some "Other" UAPs appear to have significant underwater capabilities including being able to enter and pass through the sea at great speed, there would appear to be no great requirement for them to show themselves unless they wanted to, or they wished to leave the planet.

## iv. POSSIBLE BASES ELSEWHERE

In many countries, there are not the UAP reporting networks that exist in the USA and Europe but, even so it is clear that UAPs and their bases are a worldwide phenomenon.

## b) THE BALD FACTS

In North America, there are "Other" bases at:

- Dulce
- Mount Denali in Alaska
- Mount Shasta in the Rockies
- The Rockies in Utah
- Mount Shishaldin in the Aleutian Islands
- In the sea off Marabou
- In the sea off the Bahamas
- Puerto Rico

In South & Central America, there are "Other" Bases at:

- Popocatepetl in Mexico
- In the sea off Rio de Janeiro
- In the sea south of Buenos Aires
- In the Andes in north west Argentina
- In a lake in west Brazil

In Europe, there are bases at:

- The Falkirk Triangle in Scotland
- Mount Musiné in north Italy
- Bucegi Mountains, Romania[165]

In Asia, there is not the information available to identify many bases, but there are "Other" bases at:

- Mount Kailash in Tibet
- Northern Afghanistan
- Lake Baikal in Russia
- Kola Peninsular, Russia
- The Mountains of north Japan
- Underwater in Micronesia

In the Southern Hemisphere, there are bases at:

- Pine Gap, Australia
- Blue Mountains, Australia
- Antarctica

It is likely that this is an under-estimate of the total number of "Other" bases world-wide but, even so, this does seem excessive if all they are doing is stopping us from destroying our planet with atom bombs, gathering rare earths, and studying our genetics.

What we are seeing is colonisation, with multiple "Other" species sharing the planet with us, with only their over-flights intruding on our consciousness.

## c) COLONISATION IN SPACE

Some of our wealthier inhabitants are already thinking of colonising Mars, suggesting that this is the only way to guarantee the long-term survival of the human species. I hope that they are really thinking further than this because, if some malevolent "Others" decide that they want Earth for real-estate or humans for slave labourers or food, they could probably deal with both Earth and Mars with one casual swipe.

The answer is to have as many fully autonomous colonies as possible, and to keep them small. This is what is happening on Earth. There are reputed to be over 150 generally friendly "Other" species, and the Russians are supposed to have produced a directory of 50 of these from their observations[166]. If this is so, this provides a rough estimate of the number of small colonies that could be hiding on Earth.

# Chapter 5

## WHY IS THERE NO NATIONAL OR INTERNATIONAL RESPONSE?

### a) THE USA AND ETs

The Roswell Crash occurred in 1947 CE. That same year President Truman signed the National Security Act which set up the Department of Defense, the USAF and the CIA. It is claimed that an organisation called Majestic-12[167] was set up as the same time, reporting directly to the President, to deal with all matters relating to UAPs.

In 1949 CE, Majestic's secretary, James Forrestal, died after falling from a 16th-floor window at Bethesda Naval Hospital. While officially ruled a suicide, his death was shrouded in controversy and sparked numerous conspiracy theories about a Majestic power-grab.

In 1952 CE, it is said that Washington was buzzed two weeks in a row by a cluster of UAPs. USA fighters, which were scrambled to intercept, were unable to stop them. This could be taken as a threat to the USA – "Surrender or be Hammered". The story goes that President Eisenhower was then approached by two different groups of "Others" – Nordics and Greys. He eventually made a deal with Greys. This allowed them to establish a base, and to capture earthly cattle and humans for medical experiments, provided that they returned the humans safely home. In return, they agreed to give the USA some of their technology. What he didn't realise was that they had another item on their agenda too. It soon turned out that the USA was not to be permitted to use nuclear weapons.

It may be that this is the moment when the USA surrendered its sovereignty to the ruling Greys. It is alleged that Majestic 12 works vigorously to prevent any word about UAPs from getting out, using thugs to intimidate possible eye-witnesses and their families, adopting tactics which sometimes resulted in fatalities. They instigated a policy of explaining away every reported sighting with sometimes fanciful theories, confiscating any evidence and even denying that whistle-blowers had worked where they claimed. It has now got to the state where they are trying to defend the indefensible.

It appears that the "Others" are setting the agenda, and will not allow any information to be made public until the general population is considered ready for it. The "Others" will decide when that is. Some countries around the world have decided that they are not constrained by the USA agreement with the Greys and have released everything about UAPs that they can. In the USA, the pecking order is "Others", Majestic, then the President, in that order.

## b) POSSIBLE "OTHER" INTERESTS

Some ITs, together with the earlier civilisation if it exists, are probably aware of the position regarding the Greys. The question is whether this suits them. It is likely that any earlier civilisation would already have come into contact with the Greys, and would have a better understanding of its motivations. They may be quite content to let things run their course. The situation for ITs would depend greatly on their location. Any sea-dwelling ITs would probably have come into contact with any sea-dwelling colonists over the years, and would share their concerns. For any deep-underground dwelling ITs, the position would be specific to each case.

## c) MAJESTIC'S ACTIVITIES

It has been suggested that Majestic no longer reports everything to the President, depending on whether they consider the incumbent to be to their liking (i.e. Republican or sometimes Democrat). Effectively, there has been a power-struggle which the President had lost.

It has even been suggested that, when President Kennedy proposed sharing space technology with the USSR, they were the ones who arranged his assassination. Lee Harvey Oswald would never have had the chance to kill Kennedy in Dallas, had an assassination plot in Chicago[168] succeeded three weeks earlier, a plot that has been seldom mentioned over the years. Kennedy was due to arrive in Chicago the morning of Nov 2nd to attend the Army-Air Force football game at Soldier Field and ride in a parade. Newspapers had even printed JFK's detailed travel plan from O'Hare airport to the Loop.

Although police were preparing to line the motorcade route, Secret Service officials in Chicago were deeply troubled about the visit because of two secret threats. Right-wing radical and Kennedy denouncer Thomas Vallee had arranged to be off work for JFK's visit; Vallee, an expert marksman, was arrested with an M1 rifle, a handgun and 3,000 rounds of ammo. But then there was a phone call to federal agents from a motel manager that she had seen two Cubans with several automatic rifles with telescopic sights, with an outline of the route that President Kennedy was supposed to take in Chicago that would bring him past that building. Somehow, agents bungled surveillance of those two suspected Cuban hit men. They disappeared and were never identified. No one was even sent to the room to fingerprint it or get an ID. This shows a scenario where one hit man is set up, with two others to do the job - very similar to the Dallas scenario.

Certainly USA Congress has now come to the conclusion that there had been a conspiracy to kill him and that there were probably two shooters at Dallas.

It is alleged that when President Carter took office, he was given an official briefing on UAPs and "Others" and that, afterwards he was found in his office, sobbing his eyes out. Obviously, this cannot be verified.

The reverse engineering of UAPs has continued, probably now under the control of Majestic, and the USA has now become a two part state. NASA undertakes the public part of the space programme, whilst other agencies finance the private sector to undertake Majestic's space programme. NASA has to be very careful to avoid accidentally releasing anything about Majestic's activities. In these areas, the President has to do what he or she is told.

It has been reported that Majestic's operations are now so advanced that they could leave Earth behind if they wished[169]. One wonders why some billionaires are so keen to develop their own spaceships, and go to Mars.

## d) RECENT EVENTS IN THE USA SPACE PROGRAMME

It is to be hoped that Majestic is playing a very subtle game and not desperately clinging to power at the expense of humanity. The "Others" may be in control of the USA, but it may be that they are not aware of the activities of Intra Terrestrials, the more advanced human civilisation, or PUEs. Many of Majestic's actions would need to be kept out of sight of "Others". The appearance of the TR3B over centres of population worldwide, showing as many lights as possible, could be a form of miss-direction both for humanity and ETs, diverting attention from any new vehicles being developed. The USA has done this type of thing before.

The English hacker, Gary McKinnon[170] whose extensive search through USA computer networks was allegedly conducted between February 2001 CE and March 2002 CE, gained access to Excel spreadsheets on NASA computers. One was titled "Non-Terrestrial Officers." It contained names and ranks of USA Air

Force personnel who were not registered anywhere else. This is well before the creation of the USA Space Force in 2019 CE.

The spreadsheet also contained information about ship-to-ship transfers, but he'd never seen the names of these ships noted anywhere else. He says the ships were titled USSS (Unites States Space Ship). The USA authorities tried to charge him and extradite him, but the UK has finally rejected this. The charges and proposed penalties were so severe that it would appear that the USA authorities were really out to get him. Perhaps he had found something important.

First launched in 2010 CE, the Boeing-made, automated reusable space plane X-37B, has spent as long as 908 days in space at a time. Its 29 feet (9 meters) long with a wingspan of almost 15 feet (4.5 meters).

Early in the Space Shuttle programme, the USA admitted to having a cadre of 20 Manned Spaceship Engineers[171] (MSEs) who worked with NASA on classified projects such as launching spy satellites. After the 1986 Challenger accident, as NASA struggled to return the shuttle to flight, the Air Force sped up its plans to move payloads back to unmanned rockets. By 1988—the year NASA returned the shuttle to service—it was claimed that the MSE cadre had been disbanded, its members scattered to new assignments. (Of the 27 officers in the first two MSE groups, five would later become generals.) The USA Space Force was established December 20th 2019 CE when the National Defense Authorization Act was signed into law, creating the first new branch of the USA armed services in 73 years.

Given the way governments work, the announcement that something has been cancelled doesn't necessarily mean that it really has been, so there was probably some form of secret organization in operation for the whole of the 1988-2019 CE period, as suggested by the NASA spreadsheet described by Gary McKinnon. When the USA military's classified mini space shuttle X-37B[172] returned to Earth in March 2025 CE after circling the world for 434 days, this was the 7th mission of the type. The space plane had blasted into orbit from NASA's Kennedy Space Centre in December 2023 CE on a secret mission. Launched by SpaceX, it was claimed to carry no people, just military experiments.

This vehicle did not suddenly magically appear in 2019 CE when the USA Space Service was created. It first flew in 2010 CE. Many forget that earlier versions of this Boeing craft also had a secret military role. Although ostensibly a civilian program, they conducted a series of missions from 1982-1992 CE on behalf of the National Reconnaissance Office, launching a series of classified spy satellites, and were probably not sitting idle in the 1992 -2010 CE period.

## e) THE PRESENT SITUATION

It would appear that no-one is doing anything about the presence of UAPs and "Others", because there is nothing that they can do. Some of these colonies have been here since the dawn of our civilisation, and may even be the cause of it.

The Anunnaki appear to have initiated the drive to bring humans up to a level where they could be accepted by “Others”. They have now taken a back seat. Until the start of the industrial revolution, Earth was a delightful rural idyll for any colonists, but we started some serious pollution about a hundred years ago, and then we started playing with atomic weapons.

Needless to say, the various colonists on our planet were not happy with this on two accounts: they didn’t want us to destroy the planet, and they didn’t want us signalling to malevolent “Others” that we were there. The Greys then stepped in and, along with some “Others” of a similar mindset, such as the Nordics, are starting to take control. They have certainly taken steps to prevent the use of nuclear weapons, and are keeping a close eye on our other nuclear facilities, having even stepped in to mitigate the effects at Chernobyl. It remains to be seen whether they intend to do anything about the pollution we are causing.

They are probably aware of the development of the USA’s space capability, and may even be moving towards co-ordination of activities. The Pentyrch Incident on 22nd February 2016 CE, in the UK, shows that some cooperation occurs. Who else could have tipped off the UK government about the expected arrival of a hostile UAP? The USA has responded by keeping information away from its own population and, in general, other countries are going along.

Those amongst us who have been campaigning for full disclosure by governments of the existence of “Others” and UAPs, may be unaware of the colonial aspect. Whilst the general population could probably cope with an admission to the former, the knowledge that we are really a colony planet, may be harder to take.

# *Chapter 6*

## OUR GUARDIANS

a) GENERAL

The universe is a big and cruel place. We have only come into serious contact with the more benign species of "Others". There are probably also malevolent species, and the "Others" that we know have apparently been shielding us from them.

There are reports of buildings in Siberia which have been nicknamed "Cauldrons", and which seem to have the capability to destroy aerial objects such as in the Tunguska incident in 1908 CE. It is possible that the Greys have installed a worldwide network of these, and they could be powered from the mysterious pyramid inside Mount Hayes in Alaska. The oral history in Siberia speaks of these devices being used on multiple occasions in the past.

b) AERIAL BATTLES

In the past, there have been battles in the skies, which have been reported by human spectators. There is written record of a day-long battle over Nuremberg, in which many UAPs were destroyed, and another over the Baltic in 1665. Our defenders have been putting their lives on the line to preserve our planet.

It must have been a bitter pill to swallow, when we began trying to shoot them down ourselves. Nevertheless, they seem tolerant of our ignorance and, still appear determined to protect us.

As already described, a fleet of some 200 UAPs was seen in broad daylight launching from their Catalina base in 1992 CE. There may have been more launched from other bases at the same time, perhaps unseen because of it being night-time there. Given that the "Others" are generally wary of giving away their presence, and prefer to operate at night, this must have been a serious emergency. It rather sounds like a scramble for battle.

The Pentyrch incident in the UK in 2016 CE is evidence of the Grey's cooperation at government level. Reports suggest that the RAF had received prior information of an unwelcome UAP, and were waiting for it. As soon as its arrival was detected, it was attacked, and two large explosions were heard.

## c) IN SUMMARY

Yes, there are UAPs in our skies, and yes "Others" exist. Most of these UAPs are either small un-crewed reconnaissance drones or are crewed by Greys. They appear to be fighting a desperate battle to save our planet, even at risk of their own lives.

Of course, it is not just humans that they are trying to save. There are also an unknown number of "Other" colonies in our mountains and under the seas.

INDEX

# REFERENCES

---

[i] The Alien Colonisation of Earth's Waterways P188 by Debbie Ziegelmeyer, Pub UnX Media USA 2021
[2] Wikipedia
[3] Britannica
[4] Reports of the National Centre for Science Education V24 No 2
[5] RationalWiki – The Wedge of Alud
[6] Wikipedia
[7] Wikipedia
[8] Wikipedia
[9] Broadhaven UFO 1977 by Justin Tulley 2024 Amazon
[10] UFOs Down Under Australasian Encounters by Barry Watts P154, pub Pegasus Education Group, Victoria, Australia 2017
[11] UFO Insight - The Crestview Elementary School UFO Incident 1967
[12] IFL Science – The Aerial School Incident
[13] UFOs in US Airspace Hard Evidence P261 by John Scott Chace USA 2020
[14] New York Times https://www.nytimes.com/2020/04/28/us/pentagon-ufo-videos.html
[15] Blog of the American Heritage Centre – Discover History Flying Saucers. The papers of Jack Pickett 2022
[16] The Alien Colonisation of Earth's Waterways P36 by Debbie Ziegelmeyer, Pub UnX Media USA 2021
[17] Alien Contact: UFOs in European and Asian Air Space by John Scott Chace p112, Amazon 2020
[18] Something in the Sky – UFO Sightings in the UK p24 by Joshua Whittaker pub Austin Macaulay 2024
[19] Southern Illinois Tourism, Jackson County 2000
[20] Patch Tinley Park IL, UFO Sightings 2019
[21] Vice Newsletter by Nathaniel Janowitz 2023
[22] The Alien Colonisation of Earth's Waterways P184 by Debbie Ziegelmeyer, Pub UnX Media USA 2021
[23] Memory Cherish Palenque Astronaut
[24] SashaBlack Ancient Flier 2016
[25] Ancient Origins – Mysterious Phenomena 2013
[26] Ancient Origins – Mysterious Phenomena What became of the coneheads? By Karen Mutton 2019

[27] The Alien Agendas pp 11-21 by Richard Dolan. Pub Richard Dolan Press, New York 2020
[28] UNESCO World Heritage Convention: The List
[29] Wikipedia
[30] Britannia
[31] Scientific & Esoteric Encyclopedia of UFOs, Aliens & Exterrestrial Gods V4 p156 by Maximillien de Lafayette pub New York 2014
[32] Wikipedia
[33] Wikipedia
[34] Inf news Did aliens help Qin Shihuang to rule the world?
[35] Ancient Origins the 'Bearded God' Named Quetzalcoatl - 0014066
[36] Ancient Code, Ancient Aliens, Viracotcha 2024
[37] Wikipedia
[38] Wikipedia
[39] Mahabharata – gathertales.com The Legend of the Battle of Kurushekta
[40] Wikipedia
[41] The Times of India Mar 22 2019
[42] Operation Disclosure Exopolitics: Ancient Alien Weapons
[43] The Giza Power Plant by Christopher Dunn. Amazon 1998
[44] The Giza Pyramids Alignment Guide by Czeszkiewcz Amazon 2022
[45] Wikipedia
[46] Wikipedia
[47] UNESCO World Heritage Convention: The List
[48] The culturetrip.com The Story Behind the Underground Cities of Turkey
[49] Nature NPJ Heritage Science 2018
[50] Wikipedia
[51] Something in the Sky. UFO Sightings across the UK p11 by Joshua Whittaker Pub Austin Macaulay 2024
[52] The Soviet UFO Files P16 by Paul Stonehill Pub Bramley Books 1988
[53] State Museum of Berlin The air battle of Stralsund.
[54] Majic Eyes Only p48 by Ryan Wood pub Wood Enterprises USA 2024
[55] Majic Eyes Only p59 by Ryan Wood pub Wood Enterprises USA 2024
[56] Crystallinks – Project Haunebu
[57] Alien Contact: UFOs in European and Asian Air Space by John Scott Chace p262, Amazon 2020
[58] Alien Contact UFOs in European & Asian Air Space P22 Amazon John Scott Chace 2020

---

[59] UFOs in US Airspace Hard Evidence P377 by John Scott Chace USA 2020
[60] Scientific & Esoteric Encyclopedia of UFOs, Aliens & Exterrestrial Gods V4 p56 by Maximillien de Lafayette pub New York 2014
[61] The Flatwoods UFO Monster The Skeptical Enquirer November 2000
[62] Shoot Them Down The Flying Saucer Air Wars of 1952 by Feschino 2000
[63] UFOs in US Airspace Hard Evidence P403 by John Scott Chace USA 2020
[64] UFOs in US Airspace Hard Evidence P403 by John Scott Chace USA 2020
[65] Daily Mail President Eisenhower had three secret meetings with aliens, former Pentagon consultant claims 16th Feb 2012
[66] The Alien Colonisation of Earth's Waterways P184 by Debbie Ziegelmeyer, Pub UnX Media USA 2021
[67] The Alien Colonisation of Earth's Waterways P189 by Debbie Ziegelmeyer, Pub UnX Media USA 2021
[68] Trinity the Best Kept Secret by Vallee and Harris 2021
[69] UFOs in US Airspace Hard Evidence P195 by John Scott Chace USA 2020
[70] Majic Eyes Only p186 by Ryan Wood pub Wood Enterprises USA 2024
[71] Yesterday's America Area 51 History: Secrets Unveiled
[72] Alien Contact: UFOs in European and Asian Air Space by John Scott Chace p145, Amazon 2020
[73] Unsolved Mysteries Kecksberg UFO
[74] Underground Alien Bases P15 by Jade Summers Self Published 2024
[75] UFOs in US Airspace Hard Evidence P283 by John Scott Chace USA 2020
[76] Dark Files Bk1 P95 by Michael Schratt Amazon GB 2020
[77] The Alien Colonisation of Earth's Waterways P25 by Debbie Ziegelmeyer, Pub UnX Media USA 2021
[78] UFOs in US Airspace Hard Evidence P439 by John Scott Chace USA 2020
[79] UFOs in US Airspace Hard Evidence P439 by John Scott Chace USA 2020
[80] The Alien Colonisation of Earth's Waterways P28 by Debbie Ziegelmeyer, Pub UnX Media USA 2021
[81] The Soviet UFO Files P28 by Paul Stonehill Pub Bramley Books 1988
[82] The Soviet UFO Files P43 by Paul Stonehill Pub Bramley Books 1988
[83] The Soviet UFO Files P39 by Paul Stonehill Pub Bramley Books 1988
[84] The Soviet UFO Files P80 by Paul Stonehill Pub Bramley Books 1988
[85] The Soviet UFO Files P86 by Paul Stonehill Pub Bramley Books 1988
[86] The Soviet UFO Files P92 by Paul Stonehill Pub Bramley Books 1988
[87] The Soviet UFO Files P68 by Paul Stonehill Pub Bramley Books 1988
[88] The Soviet UFO Files P66 by Paul Stonehill Pub Bramley Books 1988

---

[89] UFO Insight 2017
[90] The Soviet UFO Files P70 by Paul Stonehill Pub Bramley Books 1988
[91] Something in the Sky. UFO Sightings across the UK p20 by Joshua Whittaker Pub Austin Macaulay 2024
[92] Broadhaven UFO 1977 by Justin Tulley 2024 Amazon
[93] The Dechmont Wood UFO Incident by Malcolm Robinson Amazon 2019
[94] Wikipedia
[95] Encounter in Rendlesham Forest by Nick Pope, pub Thomas Dunne Books 2014
[96] Something in the Sky. UFO Sightings across the UK p24 by Joshua Whittaker Pub Austin Macaulay 2024
[97] BBC News Channel 20th June 2008
[98] https://herald.wales/national-news/pentyrch-the-greatest-ufo-cover-up-of-the-21st-century/
[99] Wales Online UK News March 2024
[100] Alien Contact: UFOs in European and Asian Air Space by John Scott Chace p59, Amazon 2020
[101] Alien Contact: UFOs in European and Asian Air Space by John Scott Chace p85, Amazon 2020
[102] Alien Contact: UFOs in European and Asian Air Space by John Scott Chace p85, Amazon 2020
[103] Alien Contact: UFOs in European and Asian Air Space by John Scott Chace p61, Amazon 2020
[104] Nomanzone: UFO Sightings in Greece 2024
[105] Alien Contact: UFOs in European and Asian Air Space by John Scott Chace p80, Amazon 2020
[106] Sosacuelt.fi Mysteries in the sky
[107] The Alien Colonisation of Earth's Waterways P167 by Debbie Ziegelmeyer, Pub UnX Media USA 2021
[108] Stargates P98 by Betsey Lewis Pub Ingram Content group UK 2021
[109] The Alien Colonisation of Earth's Waterways P174 by Debbie Ziegelmeyer, Pub UnX Media USA 2021
[110] Alien Base page 313 by Timothy Good Pub Century London 1988
[111] Alien Base page 395 by Timothy Good Pub Century London 1988
[112] International Business Times 426515 2013
[113] Majic Eyes Only p278 by Ryan Wood pub Wood Enterprises USA 2024
[114] Unsolved Mysteries Mexico UFO

[115] Wired  Mexican Airforce Film 2004-5
[116] UFOInsight  UFOs Dr Enrico Bossa 2025 by Marcus Lowth
[117] Alien Base page 171 by Timothy Good Pub Century London 1988
[118] Alien Base page 171 by Timothy Good Pub Century London 1988
[119] Alien Base page 198 by Timothy Good Pub Century London 1988
[120] UFOs in Central And South America Airspace p169 by John Scott Chace 2020
[121] Wikipedia
[122] New York Post  Weird but True  2022 Aliens in Brazil
[123] Alien Base page 381 by Timothy Good Pub Century London 1988
[124] Media.com  Christina Gomez Secret Argentinan UFO Cases 2024
[125] The Dragon's Triangle by Charles Berlitz pub Winwood Press New York 1989
[126] Alien Contact: UFOs in European and Asian Air Space by John Scott Chace p71, Amazon 2020
[127] Alien Contact: UFOs in European and Asian Air Space by John Scott Chace p217, Amazon 2020
[128] Alien Contact: UFOs in European and Asian Air Space by John Scott Chace p206, Amazon 2020
[129] Theufodatabase incidents The Kofu Incident
[130] The Dragon's Triangle by Charles Berlitz p32 pub Winwood Press New York 1989
[131] The Dragon's Triangle by Charles Berlitz p130 pub Winwood Press New York
[132] Britannica
[133] News on Japan article 145271
[134] UFOs Down Under Australasian Encounters by Barry Watts P81, pub Pegasus Education Group, Victoria, Australia 2017
[135] UFOs Down Under Australasian Encounters by Barry Watts P154, pub Pegasus Education Group, Victoria, Australia 2017
[136] UFOs Down Under Australasian Encounters by Barry Watts P57, pub Pegasus Education Group, Victoria, Australia 2017
[137] News.com.au  April 27  2013
[138] Project Raintall.  The Secret History of Pine Gap by Tom Gilling pub Allen & Unwin 2019
[139] UFO Insight Pine Gap
[140] UFO Insight Pine Gap
[141] UFO Insight Pine Gap
[142] World Press Tianjin Airport, 2024

[143] Aliens Amongst Us, The UFO Secrets of Africa by Adrienne Jaffery pub Ingram Content Group, UK 2024
[144] Aliens Amongst Us, The UFO Secrets of Africa by Adrienne Jaffery pub Ingram Content Group, UK 2024
[145] Alien Base page 30 by Timothy Good Pub Century London 1988
[146] Alien Base page 41 by Timothy Good Pub Century London 1988
[147] Alien Base page 57 by Timothy Good Pub Century London 1988
[148] Alien Base page 97 by Timothy Good Pub Century London 1988
[149] Alien Base page 100 by Timothy Good Pub Century London 1988
[150] Alien Base page 178 by Timothy Good Pub Century London 1988
[151] Stargates P125 by Betsey Lewis pub Ingram Content group UK 2021
[152] BBC Travel 2022
[153] University of Nottingham: Pavlopetri 2013
[154] Harvard University Excavations at the Submerged Neolithic site of Atlit Yam, off the Carmel Coast of Israel
[155] Mahabalipuram Submerged Ruins 2022
[156] USA Today 80470829007
[157] The Living Moon Particle Beam weapon 2004
[158] https://www.msn.com/en-au/news/other/top-secret-anti-gravity-spy-plane-tr3b-black-manta/vi-AA1w0LhM
[159] Stargates P27 by Betsey Lewis Pub Ingram Content group UK 2021
[160] Vocal.Media/History/Uncovering the Mystery of the Dark Pyramid
[161] Irish News UFO footage in Alaska 2024
[162] Alien Base page 224 by Timothy Good Pub Century London 1988
[163] Alien Base page 224 by Timothy Good Pub Century London 1988
[164] The Alien Colonisation of Earth's Waterways P151 by Debbie Ziegelmeyer, Pub UnX Media USA 2021
[165] The Secret Alien Base of Bucegi Mountains by Aramescu Florin, UK
[166] The Book of Alien Races by Gil Carlson Blue Planet Press 2017
[167] The Alien Colonisation of Earth's Waterways P181 by Debbie Ziegelmeyer, Pub UnX Media USA 2021
[168] The Plot to Kill President Kennedy in Chicago by Palamara Amazon 2024
[169] The Secret Space Program and Break-Away Civilisation by Richard Dolan
[170] Cibernews:Tech 2023
[171] Wikipedia
[172] Boeing Defence Autonomous Systems

www.ingramcontent.com/pod-product-compliance
Lightning Source LLC
Chambersburg PA
CBHW040758220326
41597CB00029BB/4988